T0088149

THE
BIGGLE SWINE BOOK

MUCH OLD AND MORE NEW HOG
KNOWLEDGE, ARRANGED IN ALTERNATE
STREAKS OF FAT AND LEAN

BY

JACOB BIGGLE

———

ILLUSTRATED

———

*" The pig, the rent payer of Europe, the
mortgage lifter of America."*

———

Skyhorse Publishing

Skyhorse Publishing books may be purchased in bulk at special discounts for sales promotion, corporate gifts, fund-raising, or educational purposes. Special editions can also be created to specifications. For details, contact the Special Sales Department, Skyhorse Publishing, 307 West 36th Street, 11th Floor, New York, NY 10018 or info@skyhorsepublishing.com.

Skyhorse® and Skyhorse Publishing® are registered trademarks of Skyhorse Publishing, Inc.®, a Delaware corporation.

Visit our website at www.skyhorsepublishing.com.

10 9 8 7 6 5 4 3 2 1

Library of Congress Cataloging-in-Publication Data is available on file.
ISBN: 978-1-62636-148-5

Printed in China

CONTENTS.

ILLUSTRATIONS OF BREEDS.

WILD BOAR.

PREFACE.

 Hog husbandry is undergoing changes. New feeding methods have come into vogue ; methods based on a better understanding of foods and food effects. New breeds of hogs have come into existence ; breeds resulting from intelligent and persistent effort to adapt animal to locality, and to the foods of that locality, and to special market requirements.

Experience has heretofore been the main guide, but science now comes to the swineherd's aid. Experience could only say that certain results would follow certain causes, but science now explains the causes. This is equally true of breeding, feeding, and the treatment of diseases, and there is less blind groping than formerly.

Railroad development, the establishment of great abattoirs or slaughtering establishments, better export facilities, etc., have produced marked effects upon the hog business in America during recent decades.

In the preparation of pork products for market, there is, I think, a distinct tendency visible toward neater and more attractive packages, and also an increasing demand for lean or marbled meats rather than for excessively fat meats.

The lard hog has been challenged by the bacon hog, and the indications are that the pig of the future will be killed younger and smaller than the pig of the past.

There seems to be an increasing willingness to regard the hog as a cleanly animal, capable of living apart from knee-deep filth, and able to drink pure water and to eat grass.

As to swine diseases, great progress has been made in their study. Hostile bacteria are gradually coming under control, but it is very evident that epidemics are more easily prevented than cured. Cleanliness is fully warranted by all economic considerations. Much space is devoted in the following pages to a review of hog cholera and other swine ailments.

I give three chapters to the subject of feeding because of its prime importance in profitable hog husbandry. The proper balancing of rations is now quite fully understood by leading live stock men everywhere, but there are still thousands of people who are wasting food in the pig sty and in the barn. It is worth while to give thought and time to the study of balanced rations, for the theory applies with equal force to all the live stock on the farm, and feeding tables and analyses are now within the reach of everybody, free of cost. It will be the fault of the American people themselves if they do not henceforth feed their farm animals wisely and economically, for the Government and the experiment stations have placed ample data freely at their disposal. I hope this little book will prove helpful in the same lines.

JACOB BIGGLE.

Chart Showing External Hog.

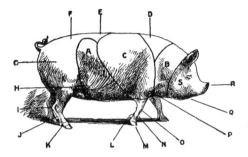

A	Abdominal Region	K	Pastern-joint
B	Neck	L	Ergots or Rudimentary Claws (Front)
C	Chest or Thoracic Region	M	Claws
D	Withers	N	Front Cannon Bone
E	Back	O	Knee-joint
F	Croup	P	Shoulder-joint
G	Hip-joint	Q	Jowl
H	Stifle-joint	R	Snout
I	Hock	S	Head
J	Cannon Bone		

Diagram Showing the Different Cuts of Meat.

KNEE DEEP IN PASTURE—LAUGH AND GROW FAT.

CHAPTER I

PIG FIGURES.

A great business this: millions in it, literally.—Tim.

A quarter of a century ago there were about thirty million hogs in the United States. At the present time there are nearly fifty million pigs in this land of corn.

Iowa has more than seven million, and Missouri and Illinois each have more than three million hogs.

The average farm value of the swine of the United States, per head, is placed by the U. S. Department of Agriculture at $5.99. The highest valuations are to be found in the New England and Middle states, varying from $9.45 in Maine up to $12 in Connecticut. Iowa is credited with an average price of $6.71 and with a total hog valuation (U. S. Yearbook, 1904) of more than $48,000,000. This enormous total valuation is well nigh double that of Illinois, the next competing state, which is placed at something more than $25,000,000, but with an average of only $6.74 per animal. Ohio and Indiana have hog valuations of more than $15,000,000 each.

In American live stock interests, cows and cattle rank first in value, horses and mules second, hogs third and sheep fourth, and yet our export of pork products is $105,146,945, which is more than the export of horses, cattle and their products combined.

A great deal is said about the "corn belt" of the central West, but it is an error to suppose that all the pigs of the country are produced there. The fact is that the Southern states produce great numbers.

The U. S. Government, through the Bureau of Animal Industry, now makes a scientific and careful inspection of all meats intended for interstate or foreign commerce. "The sanitary value of the system," says a recent report, "is beyond computation. It protects life and health. Inspection will become so general and so perfect that not a single pound of unwholesome meat will find its way from the United States to foreign markets, nor will any be sold at home which does not carry a certificate of inspection." During a single year, recently, the inspectors examined, microscopically, 979,380 specimens of pork, either whole carcasses or pieces, and found only a little more than one per cent. of samples containing trichinæ.

There is at present no official inspection of home consumed pork, or of pork intended for sale in the markets of the state where the hogs are grown and fattened.

It has been asserted, though I cannot say how truly, that there are more swine in the United States than in all the rest of the world combined. This is doubtful, though it is quite probable that no other part of the world produces such a bountiful supply of excellent hog foods as the United States.

CHAPTER II

MANY HOGS OF MANY KINDS.

There is no best breed of hogs, but several of the breeds are mighty near the mark.—John Tucker.

The breeds of hogs raised in America to-day are Poland China, Berkshire, Chester White, Duroc Jersey, Tamworth, Yorkshire, Cheshire, Victoria, Essex, Suffolk, etc., and their proportion is in somewhat the order named.

The Poland China, Berkshire, Essex and Hampshire or Thin Rind breeds are black, or black with white markings.

The Chester White, Yorkshire, Cheshire, Victoria and Suffolk breeds are white.

The Duroc Jersey and Tamworth breeds are red or brown.

The Berkshire, Yorkshire, Tamworth, Suffolk, Essex and Hampshire or Thin Rind are of English origin.

Of American origin are the Poland China, Duroc Jersey, Chester White, Cheshire and Victoria.

Brief and necessarily incomplete descriptions of the several breeds now in favor in the United States will be found in this chapter. It is natural for every enthusiastic hog man to think his breed is the best of all, and worthy of minute detailed description. Yet breed is much a matter of choice, and the best hog is the one which gives the greatest return for food consumed. Different locations seem to demand different types, and new conditions actually develop new strains.

Possibly it is true that there are more American votes to-day for the Poland China breed than for any other, with Berkshires and Chester Whites and Duroc Jerseys in next highest esteem; but it is much a question of fancy. New England, for instance, seems to think well of Small Yorkshires.

POLAND CHINAS. This admirable and widely popular breed of hogs "largely originated," as one

POLAND CHINA SOW, TEN MONTHS OLD.

account puts it, "in southwestern Ohio." It had its foundation in various crossings made by stockmen in Warren and Butler counties, between the years 1816 and 1842, though the present breed name was not finally adopted until 1872. The present fancy calls for pure black, or nearly so, with white points; that is, white feet, white tip of tail and white about nose. Often there are sheets or patches of white on the body, but the white color is being gradually bred out. The size of the Poland China is large. In shape and form it differs from the Chester White in being shorter in the legs, broader in the back, with larger and heavier hams according to size of carcass. The old-fashioned Poland China had drooping ears, but the modern type has a thin ear which tips nicely at the point rather than droops. In disposition it is very docile. It easily fattens at any age, and is in all respects an excellent breed. The name is unfortunate, for there is no Polish breed.

The history of the Poland China hog is another of many proofs of Prof. Brewer's conclusion, after studying the origin of many breeds of horses, cattle, sheep and swine, viz., that no breed has ever been established by the crossing of only two breeds, but by the crossing of three or more distinct breeds. As early as 1816 there were used in Warren and Butler counties, Ohio, the Russian, the Byfield and the China, all mostly white, and their produce was known as the Warren County hog. Its size was very large and it was usually white. In 1835 importations of Berkshires were made by a packer, M. Beach, and used. In 1840 another Cincinnati packer imported Irish Graziers, a white hog; after which no other blood was introduced, unless we so consider the red boar, owned by a Polander, in Warren county, and his produce was called Poland pigs, and were usually red or spotted.

Prior to the National Swine Breeders' Convention in 1872, when the name Magee was given the breed, it had many names, such as Dicks Creek,

Gregory Creek, Miami, Butler County, Warren County, and further west it was known as the Moore hog, since he had taken hogs from his old home in Warren county, to Canton, Illinois, and became a prominent breeder. This name, Magee, was too personal and did not tell the truth as to origin. So at the next meeting of the National Swine Breeders' Convention the name Poland China was adopted, not because it tells anything of origin, but because it so effectually shut out any clue to place of origin or the name of any individual who had a prominent place as breeder or disseminator of the breed.

The name Miami was contended for as historical and as truthfully telling the place of its origin after the English custom, as seen in the names Berkshire, Yorkshire and so on, but it is too late now to change it. So long as the rose smells as sweet by any other name, this valuable breed feeds and sells as well with this polyglot misnomer.

BERKSHIRES. This English strain, brought to America in 1823, is in high favor here. It is likewise

PRIZE BERKSHIRE BOAR, UNDER ONE YEAR.

in favor in England, in Australia and in other parts of

the world. It is a black pig, of medium size, with a dished face. Typical specimens have white points; that is, white feet, nose and end of tail. The breed originated in the county of Berks, England, probably from crossings of the local breeds with Chinese and Neapolitan stocks. The meat of the Berkshire is in high esteem, the fat and lean being well intermingled, and the bones comparatively small. The animal grows steadily, under good treatment, to an early maturity, and is adaptable to its surroundings. With pasture and exercise the Berkshire is a good bacon hog; otherwise it may go too much to lard. Variations of this standard stock are advertised under the

A CHESTER WHITE BEAUTY FROM THE WEST.

names of Large Berkshire, Large English Berkshire, Long English Berkshire, Large Improved English Berkshire, etc.

CHESTER WHITES. This is one of the largest breeds, long and deep of body, with broad back and deep, full hams. The legs should be short, and the head also short and broad between the eyes. The face is not much dished. The ears project forward. The hair is plentiful and sometimes wavy. The breed originated in Chester county, Pennsylvania, and is now widely distributed, and is in high favor. Chester White hogs have been among the heaviest porkers ever produced. This breed has been variously modified, and is also advertised as Improved Chester White, Ohio Improved Chester White, Todd's Improved Chester White, etc.

DUROC JERSEYS. This breed, or its variations, is

DUROC JERSEY.

also on the American market under the names of Jersey Red, Duroc, Red Duroc, etc.

Prior to the establishing of a record for this breed of red hogs, they generally went by the name of Jersey Reds. But this name did not recognize the fact that in New York there was a red hog known as the Duroc. At the suggestion of Colonel Curtis, the name Duroc Jersey was given to the record, which was to receive both the Jersey Reds and Durocs for record, and truthfully indicate its origin, which was about 1850. The Durocs originated in New York and the Jersey

Reds in New Jersey, and are claimed to have for ancestors the Red Berkshire and Tamworth.

The Duroc Jersey was a large, coarse hog, had a long, deep body, broad back and heavy lopping ears. Its color varies from a cherry red to brown and even to a light yellow. Sometimes there is a little black on the lower portion of the body or on the legs, but since the establishing of a standard the appearance of black is quite objectionable. Since the breed has been introduced into the West it has changed in form, as have all breeds on a main corn diet. The bone is less and the angularity has given way to plumpness. The ear, too, is smaller, and breaks or tips instead of lopping over the eyes.

TAMWORTHS. This English breed is of recent origin, deriving its name from Tamworth, an English town adjacent to both Staffordshire and Warwickshire, in which counties it is abundant. In color the animals

AGED TAMWORTH SOW.

are red, chestnut or brown. The Tamworths are bacon hogs, noted for their large production of lean meat of especially fine quality. The sows are usually prolific, and the pigs are rapid growers and mature early. The body is massive, the head small, the ears erect, and the snout inclined to be long. The hair is long, silky and thick. In England they are popular, and not a few are now in the United States. The breed is sometimes called Red Tamworth.

YORKSHIRES. There are three more or less distinctly defined strains of Yorkshires, known respectively as Small, Middle and Large. All are white.

The Small Yorkshire is regarded as the smallest and finest of swine. It has a dished face, short snout, heavy and deep body, short legs and fine bone. Small

SMALL YORKSHIRE.

Yorkshires mature early and are very docile. They are popular in England and in some parts of America.

The Middle Yorkshires and Large Yorkshires have many points in common. They quickly mature and produce a large proportion of lean meat. They are long and deep in body, short in head, conspicuously dished face, and strong in bone. The skin is pinkish in color, with an occasional bluish spot, and the hair is white, thick and soft. The ears are of good size

and point forward. Yorkshires are advertised also
under the names of Improved Yorkshires, Improved
Large Yorkshires, Large White Yorkshires, Improved

AGED MIDDLE WHITE YORKSHIRE SOW.

White Yorkshires, etc. They are prized as prolific
bacon hogs.

CHESHIRE. This excellent white breed originated
in Jefferson county, New York, and the animals used
to be occasionally called Jefferson County pigs.
Harris says that they were at first exhibited as Cheshire
and Yorkshire hogs, and afterward as Improved
Cheshires or even as Improved Yorkshires. These
facts give a pretty good idea of their origin. They
have been widely distributed, and are now known only
as Cheshires. The old English Cheshire breed was
large and coarse, but the American Cheshire is a great
improvement. The color is white, the ears small and
fine, cheek well developed, bodies rather long, good
shoulders and hams, and comparatively small bones.
The breed is a valuable one, and is popular in certain

parts of the country. There is a Cheshire register.

VICTORIAS. This white breed was originated in Indiana, within recent years, by George F. Davis. The ancestry is said to be Poland China, Chester White, Berkshire and Suffolk. The Victoria is a

VICTORIA.

larger hog than the Small Yorkshire, and more on the style of the Berkshire, but white in color and of quieter disposition. It is said to fatten more readily than the Berkshire. There is a herd register of Victorias, three volumes of which have been issued (1905).

The name Victoria was formerly applied in England to a cross between the Small Yorkshire and the Cumberland Small, but that stock now goes under the name of Small Yorkshire, and the name Victoria in America refers to the Indiana breed above described.

ESSEX. This is a small wholly black breed of English origin, having been developed in Essexshire. It has nearly or quite passed out of existence in its native country, the result of continued crossing with the Berkshire and Suffolk. (The English Suffolk is black, the American Suffolk white.) The Essex breed is still recognized in the United States.

SUFFOLK. The English Suffolk, as above stated, is black. The American Suffolk, a white pig, is believed to be a variety of the English Yorkshire, corresponding in size to the Middle Yorkshire. It has a small head, short snout, dished face, upright ears, a short neck, good length of body, fine bone, pinkish skin and soft hair. It matures early and produces excellent meat. It is, however, rather sensitive to sudden changes of temperature.

COMMON PIGS. Of course a very large proportion of the pig population of the United States belongs

PRIZE YEARLING ESSEX BOAR.

to the nondescript class ; that is, to no particular breed. Owners of such pigs cannot do better than to cross their sows with thoroughbred boars ; but a common boar should in no case be used.

Some success has followed experiments in breeding the Southern razor-backs with thoroughbred stock, the native blood being subsequently reduced to one-fourth or one-eighth. The health and energetic habits of the semi-wild animal are thus retained, and also certain desirable flavors in the meat, while the later crosses readily and cheaply fatten.

CHESHIRE YEARLING SOW.

HOG TALK.

Go many miles to get a good sire.

Choose the breed and save the feed.

Do not trade a long body for a short nose.

Do not choose a poor boar to save a dollar or two.

American and English pig names do not always correspond.

The bacon type of hogs has more stamina than the lard type.

Cheshires should dress from 500 to 600 pounds when fully grown and fattened.

A streak of lean and a streak of fat. Breed for it, feed for it, and give exercise for it.

Two breeds of swine seldom long continue in equal favor in the same neighborhood.

Small Yorkshire and Essex have been called pig-pen breeds. They are chunky, quiet and lazy.

The best breed is the one which will rear most pigs and make most and best pork on cheapest food.

Hogs with lopped ears are not so nervous as those with upright ears. As a rule they make better mothers.

For the profitable sow lay more stress on good bone, good constitution and big litters than on a number in a herd book.

The old razor-back has been driven from the pig kingdom, and the perambulating lard tub will have to follow. Mix lean and fat.

Chapter III

THE BOAR.

The boar counts for half, and a big half.—Tim.

 If we suppose that each parent exerts the same influence upon the offspring it is easy to see that the influence of the boar is precisely equal to that of all the sows combined.

Vigor of constitution is believed to depend mainly on the dam, but outward form, structure and limbs upon the sire. Fineness of bone and early maturity depend on the boar.

Boars of all breeds should be of strong build, but without coarseness. Neck and body should be short (for the particular breed), and limbs short rather than long. Such features indicate bodily vigor and easy feeding. Compactness of form is more desirable in the boar than in the sow.

Width between fore legs, with large girth behind them, denotes active heart and lungs. Straight, strong limbs and erect hoofs indicate solidity of animal frame work. Smooth skin and soft hair denote activity of the liver and general health. Add to these qualities a quiet disposition, without laziness, and the result will probably be a good sire.

In crossing two breeds a male of the smaller breed is commonly used.

During the summer season the boar should be allowed the run of a grass lot if possible, and should be fed some grain, but not kept too fat. At the breeding season he should be in a strong and vigorous condition, and from this time on through the breeding season he should be fed quite liberally of stimulating but not fattening food. Let him be rather gaining flesh during the breeding season than losing flesh.

Do not turn him loose with the sows. This I confess is more convenient, but if the experience of good breeders is worth anything it is dear convenience. Keep him in a pen alone near the sows, and when a sow is in heat allow him to serve her once, and then return him to his pen. A boar fully grown and properly fed may be allowed to serve two sows a day for several days in succession, if necessary, but this should not be continued indefinitely if the best results are to be expected. About one sow a day on an average is about the limit.

Disappointing litters not infrequently result from over-service of the male.

BRISTLES.

Avoid in-breeding.

It will pay every time to use a thoroughbred boar.

The sire should have bran or oats ; fed for vigor, not for fat.

The young boar should be trained to be driven ; it can be done.

An ungovernable boar is a great nuisance and always dangerous.

Neighbors sometimes join in buying a good, thoroughbred boar, charging fees to outsiders.

Beware of the scrub thoroughbred. Blood without quality is worse than quality without blood.

Chapter IV.

AT FARROWING TIME.

The wise breeder will give the sows proper food and sufficient exercise and then trust nature at farrowing time.—John Tucker.

 A sow that has had wise feeding during the period of pregnancy will seldom have difficulty in giving birth to her offspring.

A sow carrying pigs is engaged in a work which demands a full supply of the tissue-making or nitrogenous foods ; foods rich in protein. Besides maintaining her own life she must secrete the material for building up the bodies of perhaps half a score of little pigs and then be ready to supply them with milk. These pigs at birth will average two and one-half pounds in weight, and it is easy to see that the function of motherhood is a severe one.

Demanding nitrogenous food, such as clover, wheat bran or middlings, linseed meal, or something else rich in protein, it is not hard to understand that a sow may suffer greatly if such foods are denied her, nor is it surprising that when pregnant sows are fed almost exclusively on corn or corn-meal (which contains only one part of protein to nine or ten of carbohydrates and fat) they should be so nearly crazy for protein as to eat their own young when they are born. Such sows are literally insane and irresponsible, the result of poor feeding. The animal is starving for a certain kind of food.

In selecting a brood sow the choice should be made from a large litter. The sow should have a long body, plenty of teats, level back, straight and short legs, fine hair and a quiet disposition. Such a sow may grow so deep as to be slab-sided, but if she possesses good traits and high constitutional vigor her coarseness can be toned down by the use of a finer-

YOUNG CHESHIRE SOW—A GOOD ONE.

built thoroughbred boar. The disposition of the young, unbred sow may be quickly learned by catching and holding her.

The disposition of the sow depends largely on her treatment from pighood up till maturity. A sow that has been kindly treated will in most cases be kind and gentle, and too much emphasis cannot be placed on the importance of having a sow so gentle that at farrowing time she will allow the attendant to enter her pen (if necessary) without becoming excited. The life of the whole litter and often that of the sow depends on the assistance which may be rendered by a skilful attendant. In an experience of over twenty years I have never lost either a sow or her litter at farrowing time on account of the sow not being able to be delivered of her pigs; though I have lost many sows and litters because the sow would not quietly submit to assistance.

Sows should not farrow before reaching the age of one year, nor should boars be used before attaining the age of eight or ten months, though many breeds will mate earlier if permitted to do so. It is a common mistake to breed hogs when they are too young.

The practice of mating a small male with a large sow, which is so common, has caused the invention of various breeding boxes or crates, of which several are now on the market, arranged so that the different sized animals stand upon different levels, the height of the rear platform being adjustable.

First litters are not usually as good as succeeding ones, and two-year-old sows are better for breeding purposes than younger animals. A first-class brood sow should be profitable for a number of years—six or seven, in some cases.

The period of gestation is sixteen weeks. I have never had a sow give birth to a litter of living pigs at less than 110 days. My experience teaches me that young sows carry their pigs in nine cases out of ten from 111 to 113 days. Old sows or sows after first litter carry their pigs from 112 to 117 days, the time increasing with age.

The pigs should suck for six or eight weeks, and the mother should have a resting period of three weeks. This will make it just about possible to raise two litters per year. It is more common in the West to allow them to suck for from ten to twelve weeks. The pigs get a better start and grow more rapidly, and, it is claimed, make stronger and better hogs.

It is a question for individual breeders to determine whether to breed once a year, twice a year,

A TAMWORTH LITTER.

or three times in two years. There are instances of three litters a year, but the best intense breeding is somewhere about two litters per year ; an average of perhaps twenty-seven weeks to each litter. The sow will usually come in heat a few days after the pigs are weaned ; sometimes in three or four days, depending on the amount of milk secretion and general condition. A sow in good vigor will come in heat sooner than one which is in poor condition.

A week or two before farrowing time the sow should be put in separate quarters, apart from the other hogs. She will carry straw and make herself a nest, and will usually require no attention. But it is well to keep an eye on the event, and there are cases where help is needed, and where surgical instruments are necessary for the safe removal of the pigs. Forceps for this purpose are on the market. Sows rarely have trouble at farrowing time if the bowels are kept open.

In the early stages of gestation no special care as to food is necessary, but as the period advances there must be an increased supply of nitrogenous food.

The sow about to farrow will eagerly eat carbonaceous food, like potatoes, turnips, apples, cabbage, roots, etc. Such food, in connection with milk, wheat bran, linseed meal or other nitrogenous food, is good for her. Clean sods, charcoal, etc., seem to have the power to aid digestion, and the penned-up sow should have such things. Never feed an exclusively corn or corn-meal diet.

Laxative food, like linseed meal, serves the double purpose of keeping the bowels open and also of supplying needed nitrogen or protein.

The brood hog should not be fat, but neither should she be thin in flesh, but must be in good condition and well nourished.

A few hours before farrowing the milk always comes into the teats. Internal nourishment of the offspring has been completed, and nature now makes provision for the new order of things. Overfeeding, or feeding with heating or constipating foods, is likely to make trouble, and hence it is common practice to feed lightly at this time. Sow's milk is much richer in fat than cow's milk.

Some breeders give no food for twenty-four hours after farrowing, but it will do no harm to furnish the sow with some bran or middlings in warm water if she seems hungry or thirsty. For three days the food ration should be light. After that she should have milk, bran slop and other nutritious foods for a week or two, and then some corn and other things for variety.

Quietness and rest are more essential than food immediately after farrowing, and the sow should remain undisturbed as much as possible.

In cold weather it is sometimes necessary to cover the new arrivals with a blanket, but this need not be done if the sty is free from draughts and not too spacious. Some breeders even use a jar or bag of hot water under the blanket.

Exercise is to be encouraged, for the sake of both dam and offspring. The bed should be wholly changed and made fresh a few days after the birth of the pigs, and wet straw carefully removed at all times. Idleness and too much food and warmth sometimes cause the little pigs to contract a disease known as 'thumps.'' The cure is difficult. Exercise is the prevention.

If a sow's teat is so sore she will not let the pigs suck, cut it off and save the pigs. Sometimes by smearing it with tar the pigs will let it alone and the sow will let her young suck ; but if she will not, cut the nipple off close to the udder and the trouble is over.

It is sometimes necessary to throw a sow on her side and fasten her in that position, in order to allow her pigs to feed. In case the sow persists in her refusal to claim her pigs, they may be kept near her in a ventilated box, and fed as indicated until she accepts them. A sow may safely be kept fastened on her side all night.

It is well to teach the pigs to eat from the trough as soon as possible, which means that it costs less to feed them direct than through the teats of the dam. This gives the sow more of the season in which to rear another litter.

As soon as the sow has gotten used to the loss of her pigs she may, if the weather be warm, be put into a pasture and allowed to run there until the next farrowing time, being fed sparingly or not at all. It is preferable, however, to give her a little wheat middlings, unless she has access to clover, peas or other nitrogenous food.

VIGILANTS.

A good pig may be of any color.

Avoid feeding corn in hot weather.

Two litters a year is good practice.

Save the best sow pigs for breeders.

Do not breed young, immature sows.

Do not kill good breeding stock too early.

Keep a record of the performance of each sow.

If the sow eats the after-birth no harm will ensue.

The spaying of sows does not seem to be profitable.

Breed any month in the year, if it suits your market.

Breed coarse, well-formed dams to finer and smaller sires.

If lice are suspected on sows use grease before the juniors arrive.

Fatten the rattle-headed sow that lies on her pigs. Try another.

It is all right to turn corn into pork, but not into mere pork oil.

Separate young sows from older ones during period of pregnancy.

There is no more profitable animal on the farm than a prolific sow.

When pork is low in price is the time to increase the number of breeders.

It was a prolific sow that presented her owner with seventy-seven little pigs in five litters.

A breeding sow can be kept on about the same amount of food that it costs to winter a shote.

An old sow is apt to be sluggish, and the risks of her killing her pigs are twice as great as with a young one.

A CONTENTED DINNER PARTY.

CHAPTER V

LITTLE PIGS.

It seems to me that the juniors always do best when neither coddled, pampered, overfed nor underfed, but just have a fair chance to take care of themselves.—Dorothy Tucker.

When the pigs are twenty-four hours old let the sow out into the air and sun for a little exercise. If the weather is cold a blanket may be needed over the young things while the dam is absent. As the pigs get older let the sow's time for exercise be gradually increased.

Young pigs, especially first litters, must be jealously guarded against cold. Early litters should be born in closely-built and protected structures, though even a tight building may be so roomy as to be unsafe. In this case throw up a temporary floor or scaffold, and cover it deeply with straw, so as to make a warm compartment for the sow and her pigs. They will need such shelter until the pigs are eight weeks old.

As a rule, March first is early enough for a litter to arrive in the Northern states, especially if the sow be green or immature. September is a good time for autumn litters.

There is a wide diversity of practice in teaching the pigs to eat from the trough. Some careful breeders feed the sow in a separate compartment, away from the pigs, lest the little ones pick up scraps for which their digestions are not yet ready. Others permit the pigs to take their chances along with the mother. Others provide a small separate trough, out of reach of the sow, and feed the pigs oats, bran, soaked corn and even wheat. The wheat should be cracked if thus fed, and I should in no case use the corn alone. The muscle-making foods will make some fat, but the fat-making foods will produce little muscle. Corn has, it is true, some muscle-producing ability, but not much, and what small pigs need is food that will make muscle, bone and blood. The real demand for King Corn will come at a later stage of the operation; not in pighood.

All in-doors or cold weather farrowing demands careful shelter and separate management for each sow, so that crowding and injury may not ensue. Sows due to pig in April are more likely to have good luck with their young than sows farrowing in February or March, unless good care and shelter be given to the earlier litters.

On the plan of two litters a year it is evident that the two lots of pigs must be cared for differently, since one lot comes in spring and the other in autumn. But the summer freedom and exercise of the dam, in connection with a diet of grass and clover, may be confidently expected to produce more thrifty pigs in September than were littered in March, and it is therefore largely a question of care and management whether the fall pigs or the spring pigs make the more rapid progress in growth and development.

Summer feeding is commonly supposed to be much cheaper than winter feeding, because in cold weather a large portion of the food is burned as fuel to supply animal heat. It is only fair to say, however, that during the sucking period the September pig has more warm weather and a greater variety of food, including grass, than the March pig. Hence the fall pig just after weaning should be a cheaper and a more thrifty animal than the spring pig at the corresponding period of its growth.

The arithmetic of the question is not so hostile to winter feeding as would at first appear, for the heat-producing foods are not expensive, and care and shelter count for much. Then there is not uncommonly a better market quotation on well-rounded six-months-old pigs in March or April than in the fall, and ready cash in the early spring is a very acceptable thing.

Castration should be attended to at the age of six weeks, while the pigs are still with the sow ; and I cannot too emphatically urge that not one of the male pigs of a common litter be kept for breeding purposes, no matter how promising its appearance may be. No reliance can be placed upon the offspring of such a male, even though sired by a thoroughbred.

Spaying of the females, which consists in removing the ovaries, is not much practised in this country.

The first thing a litter of pigs will do is to fight for milk. It is sometimes necessary to cut off their sharp front teeth, to prevent damage.

Pigs should not be weaned under eight weeks old; ten is a better age; and if the sows are bred only once a year, twelve weeks old will do better still.

The process is differently performed by different breeders. I know of no better plan than to change the food of the dam from a milk-producing to a non-milk-producing basis (corn-meal and water with grass for instance), and take her away from the pigs for twenty-four hours. Then let her return, and allow them to suck. Then keep her away for two whole days, and again allow them to suck. Then make the separation final. The pigs have become shotes.

SQUEALS.

Give oats to the youngsters.

Feed up the runt for a roaster.

Have a few pigs every year to sell to neighbors.

Attempt to have pigs of only one size in the same enclosure.

Small pigs grow rapidly in a cold rain; that is, rapidly smaller.

Avoid scours by keeping things clean about trough and swill tub.

Let runts run with sow if she is not to be bred for a second litter.

Any day in the year is suitable for a pig's birthday, if it can thus meet a market requirement.

In cold weather it will pay well to give warm food to the pigs for a time after weaning them.

After a pig attains seventy-five pounds it is ready to lay on a pound or more of flesh per day, if well fed.

Get sow and pigs on the ground by the time the youngsters are three weeks old. Grass is food and medicine.

As a last resort, where the mother has insufficient milk, put a rubber nipple on a tin bottle and assist things.

Aim for a daily gain of one and one-half pounds per pig. If you do not work for it you will probably not get it.

Chapter VI.

SHOTEHOOD.

A good start with pigs is more than half the race, for a well-started pig is nearly sure to be healthy.—Tim.

WHERE IS OUR MOTHER?

The treatment of shotes, or young hogs, is a matter of moment, for profits depend upon it. Shall we feed for bacon or for lard? Is the aim an animal weighing 200 pounds or 400 pounds? Is reliance to be placed wholly on home-produced stuffs or partially upon purchased foods?

I think most of my readers must have recognized the general tendency toward smaller and lighter hogs, as compared with old-fashioned customs. Many markets will now take 200-pound animals in preference to 300-pound animals. The meat of the smaller animal is certainly better and more palatable, and there can be no doubt that it is more cheaply produced, for pigs gain in weight much more rapidly during the first six months of their growth than during the second six months.

Local conditions must govern local practices, but wherever the light hog will sell let him be thus sold.

When it comes to the food question, and economy demands the consumption of home-grown products (no matter what), it is only necessary to properly balance such rations, either by use of what is already on hand or by the purchase of whatever may be lacking. It is high time, however, for everybody to realize that corn is not a perfect food for hogs, and must not be used alone, except for finishing.

I have no notion of allowing pigs to root up new clover sod, for that is trespassing on next year's food supply, but otherwise I am in some doubt about the necessity for rings. It is the nature of the beast to root, and perhaps it can be broken up in no manner except by the use of a piece of iron in the snout. But I am forced to think that perhaps the pig roots in the ground to satisfy certain cravings for food which would otherwise remain unsatisfied. A smoothing harrow and some grass seed will repair damage caused by rooting, especially if there is more than one pasture, so as to make it possible for the grass to grow. Each swine raiser must decide for himself whether or not to use rings. My western friend tells me that rings are never necessary in the noses of hogs fed salt and charcoal ; that a pig roots simply because there is a lack of phosphoric acid for bone growth.

There are both iron and copper rings on the market, but I do not like the latter. The ring must not be set in so deeply as to wound the bone, and never through the partition between the nostrils. In my opinion much evil has come from putting rings in the snouts of hogs.

There is perhaps no better system of hog pasturing than a series of long narrow fields. Here the hogs may eat grass, and the cultivation of forage crops in the unoccupied enclosures makes it easy to feed by simply cutting the stuff and throwing it over the fence. The narrow plats thus grow rich quite rapidly, and produce more and more pig feed. Hogs can be successfully grown in pens, but pasturage is surely better. A shaded enclosure, like an orchard is an excellent place for pigs.

Water should at all times be accessible to hogs ; preferably running water, unless the stream comes from farms where hogs are kept. Disease is likely to follow such streams. The hog is a clean animal, if given a fair chance.

The sleeping apartments should not be neglected, or they will become foul and unwholesome. Sunshine is the greatest microbe killer ; therefore let sunshine into every pig pen.

Penned pigs need a good scratching post. Get a rough log and fasten it securely in the pen, as shown in the illustration. It will be popular and it will pay. The sheds should be cleaned out frequently, both in winter and in summer.

Pigs do best in small lots ; not more than five in a nest. It is much easier to secure fair play at feeding time with a small number than with a large number. Large herds should be divided for sleeping purposes and for feeding, and pregnant sows must be looked after that they are not too much knocked about.

There will be but few sick hogs if dry, warm, clean sleeping quarters are always available. Growing pigs will not harm themselves by overeating if they can obtain sufficient exercise.

The production of lean meat is partly a matter of breed and partly a matter of food and exercise. Some breeds are known as bacon hogs, and these seem to produce much lean meat. The nitrogenous foods distinctly favor lean meat, and exercise operates in the same direction. Lean meat is muscle.

To produce lean meat practically, I give the animals a large pasture field, and allow them to eat grass and to root. I feed skim-milk and bran or middlings, and keep them there until the approach of cold weather, when they must, of course, have access to warm quarters. A little corn at the last is keenly relished by the pigs, and does not excessively increase the fat, if fed for only a few weeks before butchery. Such hogs would, of course, lay on fat very rapidly under a long-continued corn diet.

PROMISES.

Bristles denote a coarse skin.

A wet pen will make a lame hog.

The currycomb will do no harm.

Black teeth do not indicate disease.

Shift the hog pasture every year or two.

Give a hoggish hog a separate apartment.

DINNER TIME.   The hog is not responsible for poor fences.

Doctoring cannot take the place of cleanliness.

Swine, like foolish men, never back down when they are wrong.

The proper development of the pig is lean first and fat afterward.

Chicken-eating hogs need more wheat middlings, clover or skim-milk and less corn.

Chapter VII.

THE PIGGERY.

One sow and one litter in each enclosure is the ideal number.—Tim.

The ground where the piggery is to be located should be high and dry, so that the rains will wash away all filth. If the ground is well shaded and well watered, so much the better, but by no means should it be located on a stream which flows across other land, as the danger of disease is thus greatly increased. A spring located on ground over which you have entire control would be all right, but statistics show that streams are

LOOK OUT FOR CHOLERA HERE.

the most potent agency in the distribution of hog cholera germs. In the absence of a spring, water can be cheaply and conveniently furnished with a tank placed on some elevation and pipes to carry the water where needed. Windmills are now so cheap and so effective that the matter of supplying water in this manner can be accomplished at little cost. In the absence of natural shade, artificial shade should be constructed by setting crotches in the ground and laying poles or rails across and then covering with a good roofing of straw. It should be located on the north side of a fence or hedge and should not be too high. Dirt is the best floor that can possibly be used for a shed of this kind. In hot weather the hogs delight to lie on the ground and, if allowed to do so, they rarely seem to suffer much from heat. Great care should be used not to expose hogs to the sun and worry when the weather is hot, for no animal will die so quickly from heat as the hog.

For pasturing hogs during the fattening process, a good rule is to allow an acre of ground for every five hogs, letting them all run in one pasture. If raising pigs is the object, the pasture should be divided into lots, of about an acre each, with pig-tight fencing. This enclosure will be large enough for two sows and their litters, and not more than two sows and litters should be kept in the same enclosure. When several litters are allowed to run together, the strong rob the weak in spite of anything I was ever able to do. But when kept separate, all feed alike and grow alike. If I had room for only two litters I would raise but two litters. Two litters well cared for will make more money than four litters poorly cared for.

Where pigs run in a pasture or orchard, the shelter

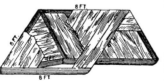 for the farrowing sow, shown herewith, has been found highly satisfactory. A horse can be hitched to it and take it to any part of the field. In the illustration the boards are cut away on one side, showing the interior. It should be made eight by four feet with

a sharp peaked roof. The runners should be made of
two-inch plank, eight feet long and two inches wide,
with holes for a chain in each front end, set four feet
apart and fastened together at the back with a piece
two by three, four feet long. This home will make ample room for a sow and her litter and can be placed
where convenient, facing the south. It will also accommodate three good-sized sows before farrowing
and keep them comfortable. Around the sides of this
home, or any other pen used by a farrowing sow,
should be fastened a strip nine inches wide and nine
inches from the floor as a protection for the little pigs.
This arrangement will prevent them being crushed
when the sow lies down, as they can escape under this
slat and crawl out at either end.

Another rather more pretentious western farrowing pen, as here shown, comes to me from Ohio. It is

six by seven feet, built on three two
by three oak scantling and is six feet
high in front and two at the rear, giving the roof a steep pitch. Grooved
pine boards are best for sides and ends, with a door
two and one-half feet wide. Place the shelter so that
the door will face the south. Put a clevis in the middle
runner so that the pen can be moved. Block the runners up so that they will not rot off. Keep the doors
closed a few days after the pigs are farrowed, according to the weather.

A convenient portable pen
for a small pig and well suited
for getting the pig out of doors

on a village place is shown herewith. It has no bottom.

The covered end protects the occupant from sun and rain. Two wheels made from a plank are screwed at one end, while handles are placed at the other. It can be moved its length every day.

Many cheap shelters for hogs used in the Ohio  valley are made by piling the straw from the stacker over and around a simple frame-work. The illustration shows a type of this cheap but effective shelter. It would be better with the sides boarded up and straw piled all over it except the front.

For a permanent hog pasture sow four quarts of clover and two bushels of blue grass where this grass does well. If not blue grass then orchard grass, the same quantity. The clover will furnish feed for the first season while the other grasses, which are slower to start, are coming up. The clover will usually die out the second year, but the blue grass or the orchard grass will hold on for years if allowed to get a start before winter sets in, so as to cover the crowns of the plants. The quantity of seed named is where the land is in excellent order, and if it is not, double the amount should be used to insure a thick sward.

Chapter VIII.

THE PIGGERY—*Continued.*

Don't give the hogs the sunny side of a wire fence for shelter, nor yet put them in a little four by six pen. A well planned hog house will pay.—John Tucker.

When building a more pretentious and permanent pig pen, the following general suggestions will be helpful. Select a dry spot where there will be natural drainage, away from the house and other farm buildings, and place the building so that it will open to the south or southeast, and far enough away from the house to avoid any bad odors reaching there. No stock enjoy a sheltered place where they can bask in the sun more than swine. Both roof and floor should be tight, warm and dry. To be shut up in a little, damp, nasty pen on a plank floor or on stones or in the mud with a wet or filthy bed is not conducive to health. While everything is warm and tight, do not overlook plenty of well arranged ventilation.

My old pen was floored with oak planks, but in my new pen I have tried a cement floor for the feeding pen and entries ; of course, the runways back of each pen are not cemented. It is not well for the hogs to sleep on a cemented floor even with a good bedding of straw, as they will work down to the cold cement, which robs them of animal heat. It takes too much corn to warm up the pigs and the cement too. I have a sleeping floor made of two inch thick planks on one side of the feeding pens with a board six inches wide along the side to keep the bedding in place. The cement floor is easily cleaned ; it does not rot and break away nor does it offer a harbor for rats as a plank floor is apt to do ; the urine and manure are not wasted. In summer the coolness of the floor is appreciated by the hogs. It should slope enough to carry off water.

Be careful not to have the floor of the feeding and sleeping pen much above the level of the yard, yet it ought to be a few inches higher so that the water will not run in. Some of my neighbors' runs are way below the level of the pen so that the hogs have to scramble up like mountain goats to get in. Have a comfortable door into the pen so that a man can get in to clean out the pens. If you have to climb over every time the pen will not get cleaned often. It should be done frequently.

Have a door between the feeding pen and the lots that can be easily shut and opened. Mine slide up and down and are worked by a pulley and a rope that extend to the entry so that it is not necessary to go into the pen to open and shut them.

There is a per cent. of gain in a good bed for hogs. When hogs squeal all night with the cold, or for lack of comfort, there is loss. Each squeal represents an ear of corn and some of them a big ear.

When hogs pile up on top of each other they are apt to get sick. The under hogs get too hot and are sure to catch cold. Either put fewer hogs together or have the bed so large and dry that it will not be necessary for its occupants to fight to get under to keep warm or on top to keep dry. However, little bedding is best

for the breeding sow. The new-born pigs get tangled
in the straw when there is too much, and they get under it and the sow lies on them. They should always
be in sight.

Here is a design for a small, inexpensive house.
The plan shows the arrangement. It is twelve and

PLAN

one-half by eight and one-half feet,
divided by the low partition *P*.
The doors are marked *DD*. The
one from the feeding room leads
out into the yard. The feed trough
is shown with the chute that
leads to it. The house is eight and
one-half feet high in front and five feet in the rear.

The eaves should project
a foot or more. The par-
tition is five feet high at
the highest part, sloping
down to six inches. *WW*
are the windows, one and

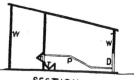

SECTION

one-half by two feet. The door leading into the yard
should be two and one-half feet high and two feet
wide.

An Ohio man sends me his plan for a combined
hog house and corn crib shown herewith. Fig. 1
is the floor plan. It is forty feet long and thirty feet
wide, exclusive of the runways marked *FFF* in the
plan. The pens are eight by ten and the entry is ten
feet wide. The outside runs are as long as you care
to make them. This house can be lengthened or
shortened to suit the requirements of the builder
by leaving off or adding pens to the plan as here

shown. *A* is the driveway entered through either
end by the sliding doors *BB*. *CCC* are the feed

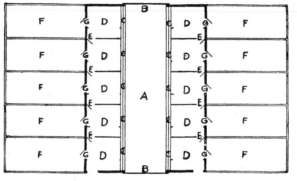

FIG. 1.

troughs. *DDD* are the pens connected by doors
with each other and by the open runs *FFF* by doors.

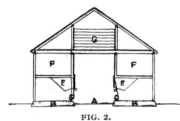

FIG. 2.

Fig. 2 shows a cross
section view. It is
twenty-four feet from
the ground to the
peak of the roof. *A* is
the doorway ; *BB* the
cement floors ; *CC* are
the feed troughs ; *DD*
the swinging doors to protect the troughs while the
swill is being poured in. The one on the right is
shown open ; the one on the left, closed. *EE* are the
chutes for the corn to come down and *FF* are the cribs.
GG, bins for bran, meal, etc. Fig. 3 is a section
lengthwise of the house. *CC* are the feed troughs ;
DD are the swinging trough doors; *EE* are the doors
to the corn crib chutes ; *FF* shows the lathing for

the crib of one and one-fourth by two inches, lath placed three-fourths of an inch apart. *HH* are

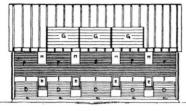

FIG. 3.

openings through which the corn is thrown into the crib from the wagon in the entry. The cement floors also form the foundation on which the building stands. This pen will accommodate from ten to twenty brood sows with their pigs, and the cribs will hold 2500 bushels of corn. The driveway may be used as a shelter for farm implements or for a cooking outfit, if desired. This plan of a pen, however, is open to one great objection in our cold northern states : if one set of pens is placed towards the south, the other set necessarily opens to the north and are consequently cold and icy.

The illustration here shown is an adaptation of the pen of a leading Wisconsin hog raiser. The floor plan

is also shown. The house is forty feet long and sixteen feet wide, with a front shed and a corn crib at the back. The cook room is twelve by sixteen feet. The well is in the shed. *BB* are doors the height of the partition and twenty inches wide hung on hinges. Partitions are three and one-half feet in height. *DD* are doors between the pens which slide up and down as do also *AA* out into the runs. *FF* are fenders made out of two by eight plank and set ten inches from the floor to protect the little pigs. They should have been shown on all sides of the pen. *GG* are the runs. The end

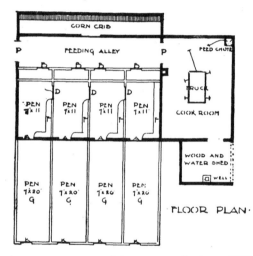

fence is permanent and has gates in it at *KK*. The side fences have a driveway in each and the interior fences are all movable. The house is seven feet high in the clear over the pen and ten feet over the cook room. The loft over the pens is used for bedding, which is let down into each pen by a little opening in the floor above. The loft is reached by a ladder. The loft over the cook room is divided into two bins. The

grain descends through the feed chute.   There is a
door *P* between the cook room and shed.   The corn
crib is three feet at the bottom and four at the top.
The slats on the outside run up and down.   The opening into the alleyway is a sliding door.   As shown
here the house has four pens, but it can be continued
on the same general plan until it has twenty.   This
pen is double boarded with wide boards and heavy
building paper in between.   The flue for the stove is
shown in the cooking room.

The piggery of a Brattleboro, Vermont, institution
which cost $8000 is here shown.   There are forty pens

A MODEL PIGGERY.

in all, twenty on each side of a central alleyway.   The
pens are eight by thirteen feet with yards eight by
twelve feet.   The entry is twelve and one-half feet wide.
The building is cemented throughout.   In the building
at the end are the steam boiler and slaughter-house.
The building accommodates 200 pigs.   Six pens near-
est the steam boiler are heated by steam for farrowing
sows.   All pens slope towards the yard.   It has all the
desirable features, such as sliding doors operated by
rope and pulleys, ventilation, swinging trough doors to
keep the pigs away while the food is being put in the

troughs, a track through the hallway, etc. Any one contemplating building a pen of this character should first see this or similar buildings.

The trough in the model piggery should be so managed that the hogs have no access to the trough into which the swill is being poured. This can be managed in several ways; perhaps the best arrangement is to have the swinging partition, as shown in the piggery on page 50. This door, which extends the whole width of the trough, is hinged at the upper side, allowing it to swing backward and forward over the trough. A heavy bolt about three feet long, with a handle on the upper end, drops down against the inner side of the trough when the door is pushed in, thus holding the door in place and keeping the pigs away from the trough when it is being filled. A sensible plan where hogs

 crowd and push is to provide the trough with partitions so that each hog has a stall, as shown in the illustration. When sows are kept in separate pastures and the number is large, as it sometimes is in the hog raising districts, the manner of getting the food to them with the least possible labor is a question worth considering. It is sometimes carried in buckets, or by a barrel provided with wheels and handles which can be wheeled from place to place as desired.

The most satisfactory way in which I have ever conveyed swill to hogs in large quantities is with a horse and a low sled. If you will try this once I do not believe you will ever again carry a bucket or wheel a

barrel, especially if you have any considerable number of hogs to feed.

COMFORTS.

Keep clean troughs.

Hogs need clean quarters as much as any domestic animal.

A pig requires plenty of water in its food but not in its bed.

Never have the chicken house over the pig pen; they want a place by themselves.

Shelter from the hot sun in summer and the cold in winter. The best summer shelter is a spreading tree.

A little pains to sun-scald the troughs, if they get sour under cover, will pay. If it be damp and cloudy scald them out with boiling water and feed a few handsful of powdered charcoal to correct acidity of the hogs' stomachs.

It won't pay to have the little pigs run out into the snow until they get large and the weather is so warm that the snow is leaving. There will be nights and mornings when the pen doors must be kept closed to hold the pigs and all the warmth in the building. The swinging doors, shown in the illustration, have been used by some and are said to work well; I have never tried them.

Provide the hogs with wallows. The wallow is the hog's bath. When he plasters himself with mud he also imprisons lice and other vermin, which he rids himself of when he scratches himself clean against a tree or a fence stake. The hog will not drink from

his wallow long after he is provided with pure water conveniently near. Wallows should be drained frequently and quicklime or diluted carbolic acid be thrown in them.

Don't lean over the fence to pour slop into the pig's trough. The fighting pigs will cause you to spill a good part of the slop, and resting your weight on your abdomen supported by a rail is not a healthful position. Pass a trough through the pen into the other trough. And if you nail a board over the top of the first trough, the pigs cannot stop it with their noses and waste the slop when it is poured in.

A CROSS-BRED FAMILY.

CHAPTER IX.

SWILL TUB AND CORN CRIB.

A pig does not eat merely to live.—Tim's Martha.

I will now discuss feeding, which is the most important detail of the business. Breed counts for much and management for more, but feeding really determines the profits.

The hog is an omnivorous animal, with an appetite for almost everything, and with ability to turn all kinds of food to account. Besides that, the hog can manage to live under conditions of the most abject misery, uncleanness and neglect. For centuries he has been regarded and treated as a sort of scavenger, and as an animal occupying the lowest position in the agricultural economy.

Now, however, things are changing for the better, and this despised farm laborer is likely to be better bred, better fed and better managed. It is now known that money can be saved by selecting the food set before the pigs, rather than by the old plan of filling them with an ill-assorted compound of stuff passing by the name of swill.

Another recent change in public opinion on the question of swine and pork is shown in the market demand for a lighter and leaner animal than was formerly in vogue. The cry for lean meat is growing louder. To my ears it sounds like a demand for better flavored and more wholesome pork, and it will no doubt result in a very largely increased consumption of this excellent meat—for pork is really good meat when not too fat.

In considering different feeding stuffs for pigs the item of cost must ever come uppermost, and I cannot assume to lay down any hard and fast lines. The feeding of whole grain, for instance, appears to me to be a very wasteful one. It does not pay to produce corn and then carry it back to the field in the shape of unbroken grains in the manure. It is also true that the miller's toll will pay for a great many lost grains of corn, while if properly fed the amount of corn in the manure can be kept down to a low point. Other stockmen may do as they find best, but I shall calculate my pig rations on the idea that all the food furnished will be digested, and not passed on to the manure pile. I find that if pigs are given time to chew their corn properly they swallow but few whole grains. It is when they are in a hurry that they bolt it whole.

The practice of fattening hogs on the undigested corn left in the manure of cattle is a very general one, and it is not uncommon for pigs to find much food in pig manure itself.

Scattering the corn is one way of making them take more time when eating. Yet I rather prefer to have all grain ground, except when fattening hogs on corn.

The dairy wastes are all excellent hog foods. Skim-milk heads the list, but is often wasted. Buttermilk is of about the same value as skim-milk. Whey is of less value, though fattening.

The wheat waste products, especially the middlings and shorts, are deservedly held in high esteem, as they are especially rich in muscle-making elements.

Linseed or flaxseed meal, including both old and new process, is very high in muscle-forming elements. The cake when ground and fed to stock makes excellent food ; it is somewhat laxative in its effect. Cottonseed meal, on the other hand, though high in its protein, is constipating in effect, and is death to hogs fed on it for a few weeks unless allowed to ferment twenty-four hours before feeding.

Corn, in its various forms, including fodder and ensilage, is perhaps the most prominent of American hog feeds. It is a grand thing in every form, a priceless boon to the agricultural public ; but alone it is not a perfect food for swine or other stock. It is highly carbonaceous ; that is, it is rich in sugar and starch, and is limited in its ability to produce lean meat, bone or milk. Its function is fattening and heating. When balanced with nitrogenous foods, and supplemented by any sufficiently bulky provender, it makes a perfect food for live stock. It will always have a prominent position in American feeding.

No better practice has yet been suggested than a hog diet consisting first of milk, then of grass and middlings, and finally of corn. I do not mean that these things should be the sole diet of the rapidly growing animal, but that they may well serve as a model of operations. Milk is well suited to building up a frame-work. A little grain, like oats, will do much good. Grass is an excellent growing food, and hogs will do well on it, especially if given a little middlings and milk, either once or twice a day. Finally, the corn will add the fat very rapidly when the pigs are to be made ready for market.

Vegetables and roots are especially useful for their effect upon the digestive system. They are an agreeable change in addition to the regular food of the hog, and also serve to keep the bowels in good order. They have what is termed a cooling effect upon the blood, which is equivalent to saying that they favor a good action of the liver. Be careful to avoid the excessive use of any one thing.

It is never a mistake to provide large hog pastures, or else to plant crops like rye, clover, sweet corn, turnips, etc., that can be cut for feed. The farmer can best determine whether to carry the provender to the pigs or the pigs to the provender.

Opinions differ in regard to the value of silage for hogs. There is a vast difference in silage itself. When the corn has ears nearly ripe the silage makes pretty good hog food, if fed moderately—say one pound per day to start with, and three pounds or four pounds per day as a maximum amount. Some feeders have pushed the amount considerably higher.

Turning hogs upon growing crops, a system known as "hogging down" the crops, is wasteful in one sense; but if it saves labor, and if the crops so treated are speedily turned under by the plow, it may at times be quite justifiable. Clover should be cut fine and steamed and meal mixed with it, for the pigs to eat it best. Soaking in water and mixing with meal does very well. It may be sprinkled with water and meal dusted over it and they will relish it. Hogs will eat clear clover hay when cut in full blossom and well cured.

Much has been said and written about feeding swill sweet or slightly soured—many good feeders advocating that swill should be mixed and allowed to stand for about twelve hours before feeding. I have studied the feeding problem as I never studied any other thing, and if there is any good reason founded on facts, either from a scientific or common-sense standpoint, that goes to prove that swill is rendered more digestible or more nutritious by being allowed to ferment, I confess I have failed to find it. I give it as my belief, founded on actual experience, that swill is not only NOT rendered more digestible or more nutritious by fermentation, but is thereby actually rendered less digestible and less nutritious. It is true that many feeders have succeeded, and succeeded well, that fed sour swill, but this by no means proves that they would not have succeeded better had they fed swill without souring. Then, by all means, let swill be fed sweet.

A word about the swill tub or milk vat : Let it be kept decently clean by at least occasional scrubbings and scaldings. A good plan is to have two. While one is in use, let the other be cleaned and stood in the sun. It is possible to have the barrel so foul as to be a positive menace to the health of the pigs. See to it.

Pack the swill barrel so that it cannot freeze. This is easily done by boxing it in roughly and packing around it closely with chaff, leaves, sawdust or charcoal. This packing should not be less than one foot on all sides and at the bottom. A heavy top box and one which can be quickly opened completes what will prove of great profit and comfort to the swine.

Keep a mixture of salt, ashes and charcoal within reach at all times. Keep troughs, quarters and food clean. There is not any reason why hogs should be more subject to disease than any other farm animals. Filth is responsible for four-fifths of the "hog cholera."

LEAKAGES.

Put the hogs in the old pasture and let them root up the grubs.

It costs but half as much to fatten a young animal as an old one.

Prepare a pig for the family roast during the first cold weather.

Pigs large or small can make good use of grass or clover in a rack.

Pigs farrowed when the sow is on grass are always healthy. This fact favors both green food and exercise.

Some farmers dump a load of light woods earth into the pen every month, and think it pays. Better turn the pigs out.

We cannot gratify a hog's ambition to possess the whole earth, but we can profitably give him a portion of the soil.

Skim-milk is by all odds the best basis on which to build up a balanced food for little pigs. It is pretty well balanced in itself.

Give plenty of pure water. Be careful that the hogs have all the water they want at night. They are apt to drink heavily before going to bed.

Chapter X.

FEEDING RATIONS.

First frame, then fat.—John Tucker.

 The arithmetic of hog feeding is simple, because the food tables are now quite complete and easily accessible. All that we need to do is to express the best practice in figures and then examine and study the figures, comparing our methods with the accepted standards. I often detect myself in error.

The following figures form a portion of the well-known feeding standards of Wolff, a German authority. These tables are now widely used in the United States:

GROWING FAT SWINE.

Age	Total Organic Matter lbs.	Protein lbs.	Carbohydrates and Fat lbs.	Fuel Value (Calories)
2 to 3 months, 50 lbs.	2.1	.38	1.50	3496
3 to 5 months, 100 lbs.	3.4	.50	2.50	5580
5 to 6 months, 125 lbs.	3.9	.54	2.96	6510
6 to 8 months, 170 lbs.	4.6	.58	3.47	7533
8 to 12 months, 250 lbs.	5.2	.62	4.05	8686

Growing animals must have a certain proportion of nitrogenous food (protein) to carbonaceous food (carbohydrates and fat). Otherwise there is a waste of food and a waste of money.   The term nutritive ratio is used to express the proper proportion.

In the German tables here given the nutritive ratio will be found simply by dividing the amount of protein into the amount of carbohydrates and fat, as the carbohydrates and fat are not stated separately, as is usually the case.   The nutritive ratio in the case of pigs two to three months old, in the German tables, is about as one to four, and would be written 1 : 4. In the case of pigs three to five months old it is just 1 : 5.   In the case of pigs eight to twelve months old it is one to six and one-half, and would be written 1: 6.5.

This all means that the proportion of nitrogenous food (protein) should rapidly decrease as the pig grows older.   At first a large proportion of protein is needed for building up the frame-work and the muscles, but later the food should be more of the nature of sugar and starch (carbohydrates), with less protein. Expressed in other words, the young pig needs milk, wheat middlings and clover, while the adult needs corn.

In tables where the carbohydrates and fat are given in separate columns it is necessary to multiply the fat by 2¼, add to carbohydrates, and divide by protein. This is because fat is 2¼ times as potent as starch and sugar in heat-making and fat-producing effects.

That is the whole story about nutritive ratio.   It is simple enough.   And yet nothing in the whole science of feeding live stock is more important than a just comprehension of food effects upon the animal system.

Yet I must frankly say that while the theory of nutritive ratio is simple enough, we are still a long distance away from exact and final knowledge in the art of feeding and fattening swine. These animals are so nearly omnivorous, and so well adapted to all kinds of treatment, that they sometimes achieve results apparently out of theoretical bounds. They have frequently been known to live and apparently thrive on both excessively narrow and excessively wide rations, but best results never follow unscientific practices.

A narrow ratio is where the carbohydrates (sugars and starches) are decreased, and a wide ratio is where they are increased, as compared to the normal amount or proportion of protein. Skim-milk and cottonseed meal are illustrations of narrow ratios, and corn and silage of wide ratios. The term nutritive ratio may be just as properly used in connection with a single article of food as with a food compounded of several ingredients :

ILLUSTRATIONS OF NARROW AND WIDE RATIOS.

	Protein lbs.	Carbohydrates lbs.	Fat lbs.	Nutritive Ratio
Skim-milk	2.94	5.24	.29	1 : 2
Cottonseed meal .	37.01	16.52	12.58	1 : 1.2
Corn	7.92	66.69	4.28	1 : 9.5
Corn silage	.56	11.79	.65	1 : 23.5

Of course the same feeding stuff will vary, particularly such a thing as silage. The development of the ears would make it a more valuable food than when cut in an immature state. Skim-milk varies widely in composition, as every farmer well knows.

It is folly on the one hand to feed nothing but skim-milk, with its narrow ratio of 1 : 2, or nothing but

corn, with its wide ratio of 1 : 10 or 1 : 12. Very
young pigs need the narrow ratio, to build up a frame-
work, and matured pigs may safely be finished on
corn ; but the most rapid growth for shotes will evidently be made on ratios varying from 1 : 4 up to 1 : 6.

It is another matter, even with this knowledge, to
maintain a perfect balance when so much promiscuous
food is fed to the hogs, and this is where experience
and good judgment count for so much.   But the
farmer who grasps the theory of the balanced ration is
certain in the long run to make cheaper and better
pork than his less intelligent neighbor who depends
solely on experience or perhaps on the advice of some-
body even less competent than himself.

It is quite possible to have well-fed pigs grow at
the average rate of a pound per day from birth to the
age of six or eight months.   Governor Hoard is quoted
as saying that the pig is at the pinnacle of profit at fifty
pounds or near that point, invariably ; and that each additional pound is slightly more expensive than its prede-
cessor.   Certain it is that young animals are more
profitably fed than old ones, and that there is a point
where feeding wholly ceases to pay.   It is also true
that the better the care the better the growth, and the
less the cost of production.

The division of labor, which constantly increases
with civilization and with improved transportation facilities, is apparent in the swine business as well as in
all other industries.   Formerly it was the custom in my
community to breed as well as to raise pigs, but now I
perceive that many of my neighbors buy more pigs
than they raise.   The pigs come as shotes weighing fifty

to 100 pounds, remain a few months and go to the market weighing about 200 pounds. Meanwhile they have had skim-milk and some grain, mostly home productions. They leave some net cash behind them, of course. I suppose this merely means that it is cheaper to carry the pig to the feed than the feed to the pig, for I live in a dairy district, and skim-milk is a by-product.

I like the sentiment of Prof. Thomas Shaw, of Minnesota, when he says that corn is to be fed all the way from the weaning period with "prudent moderation." Of course at the last it may be given with freedom, but as referring to the whole life of the hog it should not constitute as much as a half of the food. Corn is a grand food, but in the pig's middle life the ration must be carefully balanced and kept from getting too wide.

A good substitute for milk is a mess made of middlings and bran in water—two parts of middlings and one of bran. The middlings contain some flour, and the mixture is greatly relished by the hogs. The nutritive ratio is about 1 : 4.5. The amount given must depend upon the good judgment of the feeder. It is well to soak the middlings and bran some hours before feeding ; and corn may be added if it is desired to make a fatting as well as a growing ration.

In his new book on Feeds and Feeding, Prof. W. A. Henry, of Wisconsin, reduces various foods to what is called a grain basis. For instance, six pounds of skim-milk, twelve pounds of whey, etc., are considered equal to one pound of grain. To make 100 pounds of pork it requires 293 pounds of grain with young pigs as compared to over 500 pounds of grain with pigs weighing upwards of 300 pounds. Prof. Henry's tables emphasize the profit of feeding young stock. A word of caution about feeding cottonseed meal to pigs is necessary. Curtis, while at Texas Experiment Station, found that sickness appears in from six to eight weeks after cottonseed meal is added to their ration. He concludes after much study that "there is no profit whatever in feeding cottonseed in any form or cottonseed meal to hogs of any age." Later a Texas hog feeder claimed that by feeding cottonseed meal in a swill after fermenting forty-eight hours no fatality followed.

The Texas station has tested this thoroughly the past year and concludes that a light feed of cottonseed meal to pigs on pasture may be fed indefinitely, but fed in dry lots; even fermented cottonseed meal is not a safe feed. To improve the corn ration cottonseed meal may be used safely and profitably when fed on pasture or with green feed. The carcasses showed less fat and more lean meat, and the flesh firmer than when fed on corn only.

BALANCES.

Hogs are very fond of sugar beets.

It is waste to overfeed skim-milk to the pigs.

Study the difference between a growing ration and a fattening ration.

The hog, like man, is omnivorous; but a balanced ration is nevertheless needful.

Pigs consume two pounds of water with every pound of corn,—if they can get the water.

Strictly corn-fed hogs are apt to be dwarfed, weak and too fat. They are unbalanced hogs.

Vegetables and fruits are always acceptable, especially in connection with a grain or milk diet.

Chapter XI.

RECENT EXPERIMENTS.

The cheapest kind of experience is other people's experience.—Tim.

A GOOD POLAND CHINA.

To say that it does not pay to cook food for swine is not to say that the farmer's boiler has no place in the economy of feeding live stock. It pays very well, for instance, to boil small or otherwise waste potatoes in water with bran or middlings for the pigs, and to cook a hot mess for them occasionally, if only for variety; but unless the heat can be furnished very cheaply it will not pay to pursue the practice regularly. The Pennsylvania Department of Agriculture does not "know of one of the many experiments in this direction which has been continued any great length of time."

A recent winter experiment at the Indiana station in feeding whole corn and whole wheat in connection with ten to twelve pounds of separator skim-milk daily is of interest. The experiment was conducted with four lots of Chester White pigs, of the same age, for 105 days. The pigs were fed grain morning and night, and milk at noon. Those receiving whole corn gained 1.16 pounds per day, and it required 3.25 pounds of corn (with the milk) to make a pound of pork. Those receiving wheat gained 1.02 pounds per day, and it required 3.67 pounds of wheat (with the milk) to make a pound of pork. It is easy to figure out these rations arithmetically, and count the cost. The pigs were less than three months old at the beginning of the experiment.

As bearing upon the question of grinding corn or feeding it whole, the Wisconsin station found that it required 459 pounds of corn-meal or 499 pounds of whole corn to make 100 pounds of pork. In each case some middlings were used, in order to make a better ration. Practical experience favors grinding the cob and feeding it with the corn, as compared to feeding ground corn alone.

Pigs are very fond of wheat middlings, and the Wisconsin station proved economy in the use of a mixture of middlings and corn-meal, as compared to either alone. The Maine station showed wheat middlings to be far superior to wheat bran for pigs.

The Alabama station found bran to be an unsuitable food for hogs, when used in large amounts.

The Wisconsin station proved barley meal to have a high value for feeding pigs, but somewhat less than corn. Ground oats is superior in feeding value to whole oats. Oats fed with corn makes an excellent food for pigs.

In experiments with peas, at the Utah and South Dakota stations, this food was found to be superior to corn. The Alabama station found corn and cow peas to have about equal feeding values, with a superior value when combined.

The Wisconsin station found a bushel of corn to be worth four and one-half bushels of potatoes, the potatoes being cooked and fed with corn-meal.

Experiments at various stations, as summarized by Prof. Henry, of Wisconsin, showed that 615 pounds of roots would save 100 pounds of grain in fattening pigs.

The Wisconsin station fed skim-milk and corn-meal to separate lots of pigs, giving to each lot all they would eat ; also, to other pigs, skim-milk and corn-meal mixed. As between skim-milk fed alone and corn-meal fed alone, those fed on skim-milk made somewhat the larger gain.

Experience apparently demonstrates the wisdom of feeding hard-wood ashes, ground bone, charcoal, etc., where the hog diet is of necessity largely corn. Pigs thus fed have stronger bones than where they get nothing except the corn. Still in the Miami valley where they have a limestone soil, and all farmers grow their hogs on clover and blue-grass, with corn, oats and wheat middlings, it is the custom of the best breeders to keep a supply of wood ashes and salt in a dry place where the pigs can get them at will. It is common to rake up the corn cobs often, and char them, and sprinkle over this cob charcoal and ashes a little salt. This utilizes the carbon of the cobs, keeps the premises tidy and the animals more healthy.

The hog is such an omnivorous feeder, that even in the limestone blue-grass, clover and corn region, the hogs crave more ash and carbon and salt than are found in the great variety of grains, grasses, fruits, nuts and vegetables found by them. The wisdom of this practice of the Miami valley breeders and farmers, is supported by Prof. Henry of Wisconsin Experiment Station. He says : "The feeding of bone meal or hard-wood ashes to pigs otherwise confined to corn-meal diet effected a saving of 23 per cent. in the corn required for 100 pounds of gain. We further find that by feeding hard-wood ashes or bone meal to pigs fed wholly on corn, the strength of the thigh bones was about double that of pigs not allowed bone meal or ashes." Can we find any cheaper supply of bone makers than ashes? Salt and ashes and charcoal are, too, an excellent vermifuge and corrective of acidity in the stomach of heavily fed hogs.

In this chapter I introduce analyses of a number of feeding stuffs, arranged in tabular form, for use in making up rations. Fuller data will be found in all the recent Yearbooks of the U. S. Department of Agriculture. The separation of the foods into nitrogenous and carbonaceous groups is arbitrary, and merely offered for convenience, as there is no sharp dividing line. Wheat bran, middlings and shorts, for instance, though here classed as nitrogenous foods, are quite rich in carbonaceous elements also. If these tables help emphasize the facts that skim-milk is a highly nitrogenous food, and that corn is a highly carbonaceous food, they will be useful.

Various stations have reported inability to make pork on pasture alone ; alfalfa pasture is perhaps an exception.

I think that by and by farmers will all agree that certain rules apply to young animals which do not apply to older ones, and that summer treatment must be different from winter treatment, and that the whole

NITROGENOUS FOODS.

In 100 lbs. of Feeding Stuffs	Dry Matter lbs.	Protein lbs.	Carbo-hydrates lbs.	Fat lbs.
Separator skim-milk	9.4	2.94	5.24	.29
Set skim-milk	9.6	3.13	4.69	.83
Buttermilk	9.9	3.87	4.00	1.06
Wheat bran	88.5	12.01	41.23	2.87
Wheat middlings	84.0	12.79	53.15	3.40
Wheat shorts	88.2	12.22	49.98	3.83
Linseed meal (o. p.)	90.8	28.76	32.81	7.06
Linseed meal (n. p.)	89.8	27.89	36.36	2.73
Oatmeal	92.1	11.53	52.06	5.93
Pea meal	89.5	16.77	51.78	.65
Gluten feed	92.2	20.40	43.75	8.59
Brewers' grains (wet)	24.3	4.00	9.37	1.38
Brewers' grains (dry)	91.1	14.73	36.60	4.82
Rye bran	88.4	11.45	50.28	1.96
Cottonseed meal	91.8	37.01	16.52	12.58
Peanut meal	89.3	42.94	22.82	6.86
Red clover hay	84.7	6.58	35.35	1.66
Crimson clover hay	91.4	10.49	38.13	1.29
Alfalfa hay	91.6	10.58	37.33	1.38
Cow pea hay	89.3	10.79	38.40	1.51
Soja bean hay	88.7	10.78	38.72	1.54
Red clover, green	29.2	3.07	14.82	.69
Alfalfa, green	28.2	3.89	11.20	.41

CARBONACEOUS FOODS.

Corn-meal	85.0	7.01	65.20	3.25
Corn and cob meal	84.9	6.46	56.28	2.87
Ground corn and oats (eq'l parts)	88.1	7.39	61.20	3.72
Barley meal	88.1	7.36	62.88	1.96
Hominy chops	88.9	7.45	55.24	6.81
Whey	6.6	.84	4.74	.31
Corn silage	20.9	.56	11.79	.65
Corn fodder, field cured	57.8	2.48	33.38	1.15
Corn fodder, green	20.7	1.10	12.08	.37
Oat fodder, green	37.8	2.69	22.66	1.04
Rye fodder, green	23.4	2.05	14.11	.44
Timothy, green	38.4	2.28	23.71	.77
Kentucky blue grass, green . . .	34.9	3.01	19.83	.83
Hungarian grass, green	28.9	1.92	15.63	.36
Beets	13.0	1.21	8.84	.05
Potatoes	21.1	1.27	15.59	—
Turnips	9.5	.81	6.46	.11
Mangel-wurzels	9.1	1.03	5.65	.11

matter may be expressed as follows: Nitrogenous foods, like skim-milk or middlings, for all young pigs, both winter and summer; carbonaceous foods, like corn etc., for all animals, at all seasons; sparingly in summer and liberally in winter, and to fattening animals lavishly.

We must work on the scientific basis which the nutritive ratio so well suggests. Pig-feeding is better understood as the theory of nutritive ratio is better comprehended. To know the needed ratio or proportion between protein and carbohydrates in foods is to use foods economically.

SLICES OF BACON.

Give the boy a pig.

Use roots for hogs in winter. Clover and alfalfa other seasons.

The pasture must be made more of a factor in the swine business.

It sounds contradictory, but it is good advice to fatten the hogs lean.

Profit is in keeping the pound cost of production well below the pound price at selling time.

Bran makes the hog long; corn-meal makes it broad. Middlings are a better food than bran.

It has been estimated that twelve quarts of skim-milk may be converted into one pound of young pork.

At fattening time a daily bundle of clover with the corn-meal will aid digestion and improve the pork. Let the pigs grunt, but never let them squeal.

AGED DUROC JERSEY.

Chapter XII.

AN EASTERN CREAMERYMAN'S WAY

There is nothing more convincing than success, but even success can sometimes be improved upon.—John Tucker.

Here is experience; actual practice as reported by a Pennsylvania creameryman. He buys pigs weighing about 100 pounds each, keeps them sixty to ninety days, and sells them weighing nearly 200 pounds each, on the average.

The food given them is twenty-four pounds of sour skim-milk and six pounds of hominy chops per head per day. The cost of the food, which of course varies from season to season, is four to five cents per day.

The gain of weight per animal averages nearly or quite one and one-half pounds per day.

This looks like success; and the creameryman says the profits have been satisfactory.

When sour skim-milk can be purchased at five cents per 100 pounds and hominy chops at $10 per ton, the daily cost of the ration will be four and one-quarter cents. And if pork can be thus made at the rate of one and one-half pounds per day, worth five cents per pound, the daily gain will be seven and one-half cents, leaving a daily net profit of three and one-quarter cents. With 125 pigs this would mean a daily net profit of

$4.06. To this must be added the value of the manure, and from it must be deducted the cost of the labor and the item of interest on the money invested. Some allowance should also be made for accidents and losses ; but the above figures are quoted as actual results and are presumably correct.

In the case under consideration the pigs (125 in number) were bought in April and sold in June and July.

THE ANALYSIS OF TWO FOODS IS AS FOLLOWS, THE FIGURES SHOWING DIGESTIBLE FOOD INGREDIENTS PER 100 LBS. :

	Dry Matter lbs.	Protein lbs.	Carbohydrates lbs.	Fat lbs.	Fuel Value (Calories)
Skim-milk	9.4	2.94	5.24	.29	16,439
Hominy chops . .	88.9	7.45	55.24	6.81	145,345

Skim-milk varies somewhat in composition. The above figures refer to separator milk.

Hominy chops, meal or feed results from the manufacture of hominy, and contains the germ and coarse parts of the corn grain. It is quite a different food, with much narrower nutritive ratio than cornmeal, on account of the removal of the starch in the form of hominy.

A ration made of skim-milk and hominy chops, as described, would be expressed in figures as follows :

	Dry Matter lbs.	Protein lbs.	Carbohydrates lbs.	Fat lbs.	Fuel Value (Calories)
24 lbs. skim-milk . .	2.26	.71	1.26	.07	3,945
6 lbs. hominy chops	5.33	.45	3.31	.41	8,721
Total	7.59	1.16	4.57	.48	12,666

Calculating the nutritive ratio in the usual manner (multiplying fat by two and one-quarter, adding to carbohydrates, and dividing by the protein) we find that it is rather narrower than 1 : 5, which is nearly correct. This is a well-balanced ratio for the pigs when first purchased, but is too narrow for the animals when fat, as will be seen by comparison with the best feeding standards, as explained in Chapter X.

Assuming that no mistakes have been made in the above figures, either in quoting the practice of the creameryman or in the deductions which I have drawn, the question still remains, "Would there not have been an equal gain in weight on a smaller amount of food, especially if the pigs were pastured or given some bulky ration?"

The German tables, in the case of 170-pound hogs, call for only 5.2 pounds of organic matter, while this feeder gives his pigs 7.59 pounds of organic matter. This looks like a waste of food, and if it is a waste of food it is a waste of money.

A careful trial would soon settle the matter, and when I attempt to follow this man's plan I shall decrease both the skim-milk and the hominy chops, or perhaps only the former.

A daily allowance of twelve pounds of skim-milk and six pounds of hominy chops would have a nutritive ratio of about one to six, a good ratio at the finish.

If the skim-milk were reduced one-half in the latter weeks of the fattening operation there would still be an abundance of dry matter, protein and fuel value to meet the full requirements of the German tables. It would stand this way :

	Dry Matter lbs.	Protein lbs.	Carbohy-drates lbs.	Fat lbs.	Fuel Value (Calories)
12 lbs. skim-milk . .	1.13	.35	.63	.035	1,972
6 lbs. hominy chops	5.33	.45	3.31	.410	8,721
Total	6.46	.80	3.94	.445	10,693
Demanded by German standard for pigs weighing 250 pounds	5.20	.62	4.05		8,686

In the German tables the carbohydrates and fats are grouped together as 4.05 pounds. The sum of 3.94 pounds of carbohydrates and .445 pound of fat, expressed in the same way (with fat multiplied by two and one-quarter and added to carbohydrates), would be 4.94 pounds.

I do not mean to advocate a reduction in the amount of food as the pigs get older, when I suggest that the skim-milk should be cut down, but only to point out that some less nitrogenous stuff might be used in its place ; some cheap green fodder or corn, for instance. Every fraction of a cent saved in the daily cost of food is of great consequence.

NIMBLE SIXPENCES.

Skim-milk is too valuable to be wasted.

Quickly grown hogs are by all odds the most profitable.

Methods and feeds must vary with locality, but the principles of nutrition are the same in Maine and in California.

If I were that Eastern creameryman I should just go right on making money as heretofore. But I should put a little bunch of shotes in a separate enclosure, and test the German tables.

Chapter XIII.

WESTERN PRACTICES.

" There is a tax on the dog but none on the sow."

 Methods in the West differ somewhat as to circumstances and surroundings. The plan pursued by the great masses of western farmers is to have the pigs farrowed in March or April ; feed both sows and litters together till weaning time. During this time, both sow and litter have free access to a clover pasture, if possible, or in the absence of clover, blue grass pasture will answer very well. At weaning time the sows are taken away and the pigs are left in their accustomed quarters, and if they have been taught to eat corn and other kinds of feed, they will scarcely notice the absence of the sow, and will continue to thrive.

They are now liberally fed on corn and allowed the run of the pasture as before. The more careful and progressive farmers feed slops made from middlings, but by far the greater number feed nothing but corn and grass. This does very well as long as the grass lasts, but when the fall and winter season comes on, and the hog is compelled to subsist entirely on corn, it is no great wonder that he fails to make the return for the feed consumed that he should.

But it is true that the clover field and corn field furnish the cheapest feed known in the middle West. There it is the custom of progressive farmers to let pigs and sows run on clover from May to December, and after corn is mature the sows are removed to other pasture, and the spring pigs fed corn on the clover field. As the clover gets woody about the time the corn has passed the milk stage, it is cut up and thrown to the shotes, in small quantities at first. They eat the soft ears, and then chew up the stalks for the saccharine ; and by the time the stalks are too dry for them, they are ready to take the ear corn, and lay on flesh as never before.

The change from clover to new corn is gradual, and yet when on full feed of corn, the hogs will be seen every day grazing part of the time, thus balancing their ration better than we can do by mathematics.   In fact, give the hog a chance to get a variety, and with his omnivorous taste he will balance his ration better than any station feeder can do for him by feeding in a pen or dry lot.   The men who have trouble with disease after beginning to feed new corn are those who have half starved their hogs during the summer, and then began to crowd them by heavy feeding of new corn.   We have seen such men throw a whole wagon load of corn from the field out to fifty or sixty hogs, at one time, and after that was eaten they would haul out another load, and stall them again.   Is it strange that we hear of hog cholera in early fall, more than at other seasons ?

Pigs farrowed in March or April should go to market about Christmas, weighing from 275 to 300 pounds. Some may say this is too large a gain for the time, but it is not larger than has been made in hundreds of instances and can be made by any careful feeder, with the right kind of stock to begin with. This cannot be done with scrub stock and slipshod methods ; but it can be done with well-bred stock and by careful feeding.

Another and somewhat out-of-date method, which is not wholly without its advantages, is to have the pigs farrowed in April, May or June. Allow both sow and litter the run of a large pasture (the larger the better) and feed but little grain. This compels them to take abundant exercise, and beyond question produces a hog of greater constitutional vigor than can

POLAND CHINA SOW, THREE YEARS OLD. A FAVORITE
WESTERN BREED.

possibly be produced under the forcing system. A western friend, who is an excellent authority, says : "In an experience covering a period of thirty years, I never had cholera on my farm, and I raised and fed annually from 300 to 500 head. I then changed to the forcing system, and the result was I had cholera in less than three years after changing methods."

The third method in vogue in the West is that of feeding hogs with, or rather after, fattening cattle. In the good old times of thirty-five or forty years ago, cattle were fed in droves of from 100 to 200 head, with hogs to clean up the waste. The hogs were bought weighing from 100 to 200 pounds and allowed to clean up after the cattle for from three to six months, and would make a gain of nearly, if not quite, one and one-half pounds per day. From one to two hogs were allowed to each steer, and it was generally conceded that the gain on the cattle would pay for the feed consumed, leaving the hogs clear profit.

I doubt if the bacon hog will ever gain much favor in the West, for the simple reason that corn alone will not produce the so-called bacon hog. He commands

ALL OF A SIZE.

a higher price than the ordinary hog, and he should, as he costs more to produce. He has his place, but his place is not in the corn regions of the West; let him be grown in the Northeastern states or in Canada, where corn is more costly and where wheat bran middlings and other foods adapted to the production of this kind of meat are more nearly on an equality with corn as to price. Let each state or locality adopt the breed of hog best suited to the range and varieties of food most cheaply produced in that particular locality.

While wheat bran middlings, oats and other kinds of feed have their place in making up the *best* food for hogs, one thing must never be lost sight of, and that is that corn always has and probably always will form the great bulk of the hog's ration in the West, for the very obvious reason that it is more abundantly and cheaply grown than any other of the hog foods. While it is possible to import foods and feed them to the hogs at a profit, the most profitable hog farming comes from feeding the kinds of feed produced in the locality where you live.

The plan of feeding green corn to hogs has been more generally practiced, and perhaps more generally condemned by agricultural writers, than any other; and yet in the face of all that has been said and written, I have never fed any feed, either home grown or imported, with as much satisfaction as I have green corn. I have never fed anything that seemed to bring about such a marked change for the better. It seems to me that I could notice a change in less than three days. The hair begins to look glossy, the appetite seems to improve, the whole appearance is changed. I may not be as skilful a feeder as others, but I am sure I have put on as many pounds in thirty days with green corn as I ever did in double the time with any other kind of feed.

Now I am aware that I am treading on dangerous ground, for all the ills that the hog is heir to or has been cursed with (save possibly the rushing down the hillside into the sea of the five thousand) have been attributed to the feeding of green corn. Many comparatively able writers claim that it is a never failing cause of hog cholera. It is true that many hogs fed on green corn have sickened and died while being thus fed, but this by no means establishes the fact that the green corn was the cause of the disease being contracted. I am fully convinced that the only analogy there is between hog cholera and green corn is the fact that more hogs die at the season when green corn is fed than at any other.

But do not allow an unskilled or careless person to feed green corn. Let it be commenced very moderately and very carefully, and the amount gradually increased until the full feed is reached, and I would recommend that two weeks be occupied in reaching the full ration. Green corn, fed as above indicated, is as safe as any feed ; that it is as economical I am not prepared to say. I am rather inclined to think it is not.

SMALL YORKSHIRE.

Chapter XIV.

WESTERN PRACTICES—*Continued.*

WHICH ONE
WILL YOU TAKE?

There are two methods of disposing of thoroughbred hogs in the West—private and public sales. Ten or fifteen years prior to this date, by far the largest part of the hogs were sold at private sale, which had some features to recommend and some to condemn. On the whole, I believe the buyer could make a more profitable investment at private sale, for the reason that he was more deliberate and was not influenced by the excitement common at public sales. He had more time to carefully examine and had the advantage of seeing the stock in its everyday clothes, so to speak.

There was this disadvantage, however, to the buyer. The stock were often culled over quite early in the season, so that to get first choice he was compelled to make his selections before the pigs were fully developed, and it often happened that the pig which seemed to be best at the time of making the selection did not develop into the best animal later on. Then the seller was often put to the necessity of entertaining many intending purchasers and sold his pigs one or two at a time, and in this way got his money in small sums, which to some is objectionable. Thus it was that the public sale was finally inaugurated and of late years has become the most popular way of disposing of thoroughbred hogs.

The manner of holding a public sale consists in holding the entire crop of pigs, then to advertise and sell them in a single day. This has many things in its favor, with some objections. Its advantages are that all the crop of pigs are held till they are more fully developed and the buyers all have an equal chance to see and buy the best, if they are willing to pay the price. The seller has the advantage of selling all his pigs in one day and gets his money in a lump. Then he is relieved of the necessity of keeping a boarding house, as it were, for three or four months.

Prices at some of these sales have been almost fabulous. In some instances a whole herd of from fifty to sixty animals have been sold at an average of $250 each. However, it was generally believed that there was more or less deception about many of these extreme sales, but many sales that I have known were perfectly straight and honest and have averaged over $200 each. In such cases, the animals were fashionably bred and exceptionally fine individually, and were descended from popular sires. These high prices, in my opinion, have been detrimental to the business. When any strain of animals advance in price to an abnormal figure, the temptation to substitute and counterfeit is entirely too great.

While the great majority of sales have been perfectly honest, others have been just as dishonest as could be. Of late years a few men have been seeking to boom a certain strain of hogs and have done many questionable things to carry their point, and not a little to discredit the sales of honest men. Even now there is pending in the courts a suit involving the most gigantic fraud that could possibly be imagined. What the outcome of the suit will be remains to be seen. While the causes of this suit seem to be detrimental to the business, I believe in the end it will work lasting good to the business. In future men will scan as carefully the pedigree of the man selling the hog as they do the hog himself. The sales held recently have demonstrated several things. One of these is that hogs in future are going to sell for less money and will sell directly on their merits. If they have been bred by or have been passed through the hands of men of questionable character, the careful breeder is going to let them alone. It will not do to say that because a few men engaged in the business have been dishonest the business is going to be everlastingly ruined.

SOLD.

So long as men eat, so long will there be a demand for hogs, and so long as the demand for hogs continues, so long will they be bred. So long as hogs are bred, so long will there be a demand for choice breeding stock. The time is here when the boomer and the dishonest breeder must go, and when that is accomplished the honest breeder will enjoy such a period of prosperity as he has not seen in many years. The public sale will continue, but on a modified and more common-sense basis. The indiscriminate credit system must go. Both buyer and seller will do business in a business-like manner. Men will get credit if they are entitled to it, but men that could not buy a sack of flour on credit of their home grocer will not be able to buy hundreds of dollars worth of hogs in the future as they have in the past. These, with many other reforms, will be made, and in future I look for the public sale to be as it has been in the past, the favorite method of disposing of thoroughbred hogs throughout the West.

A FAMOUS BERKSHIRE.

Chapter XV.

BUTCHERING AND CURING MEATS.

A use for every product and every product to its best use.
—Tim's Martha.

 Agriculture is subject to the same economic conditions that have so profoundly affected other industrial pursuits, and on farm as well as in factory there is a marked tendency toward a further and further division of labor. Applying the principle to pork production, for instance, it practically costs but little more to kill, dress and prepare for market a thousand hogs than a hundred hogs, and hence great butchering and packing establishments have grown up in all the principal cities and railroad towns of the country, and the business is becoming centralized. In some parts of the United States home butchering will altogether cease, I suppose, but the farm will ever retain a great deal of individuality, and in all districts remote from abattoirs the work will be done by individuals, as heretofore, for years to come.

A merciful act at slaughtering time is to stun the victim with a blow in the forehead before bleeding. It facilitates sticking. To get the opportunity for such a blow, the animal must be run into a pen or chute made for the purpose. As soon as he drops or is thrown he must be turned on his back. One person holds the hind legs, another, astride of his body, holds his fore legs, pressing them backward. The sticker, standing in front, presses back his snout with the left hand and with the right thrusts in the knife, aiming for the trail. The knife should have a blade six or seven inches long, and a keen edge at the point. Before making the fatal thrust, cut a slit two or three inches long just in front of the shoulders, and in withdrawing make a slight cut upward to give free vent to the blood. Be quite sure that the animal is dead before scalding begins.

For one or two small animals the old-fashioned barrel scalder suffices. But for a pen of hogs of any considerable number or weight the modern scalder, with fire box underneath and a rack and windlass for manipulating the carcass, is indispensable. In many communities these scalders are owned by the local butcher or by an individual, and are loaned to farmers at a certain rate per day. Where this is not the case several neighbors could well afford to buy and use one in common. They are usually mounted on a two-wheel truck that can be attached to another vehicle for easy transportation from place to place.

Many cheap, home-built devices are available for hoisting the hogs into the scalding tank, and for after-

wards hanging them up, one at a time, on the poles prepared for them. The accompanying illustration will explain itself. If the scald water is not hot enough to loosen the hair and bristles, an old and good way to increase the heat is to drop hot stones into

the water. A little air-slacked lime thrown into the tank will start the bristles and hasten the operation.

To dress a hog for market involves the opening of the animal along the entire length of the under side and the careful removal of the intestines. The operation must be so well done that the final effect is pleasing and cleanly in appearance. The skin must be free from blood stains and the meat without blemish, with the fat portions as near to ivory whiteness as possible.

Here is a simple device where few hogs are butchered. *A* is a bolt pivoting together *BBB*, three poles of equal length. *CC* are strong hooks on which to hang the animal. Hook the carcass while lying flat, then by pushing on the other pole the tripod can easily be raised to an erect position. One man can hang a 500-pound hog in two minutes.

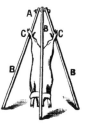

There are various methods of cutting the carcass. The diagram here shown illustrates the manner of division for mess pork. In this case the head is cut off and the carcass split through the back bone. The hams are cut round and shoulders square and the sides cut across into strips, as shown. For family use,

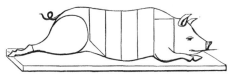

mess pork is at a discount in these days. For this reason it is a common practice now to make as much as possible of the fat into lard, and leave only the leanest cuts for pork and bacon, and to use the trimmings for

sausage and scrapple and other side dishes. The
division in this case is somewhat different. The carcass
is split on either side of the spine, leaving a strip four
inches wide. From this the fat is cut for lard, and the
lean meat and bone cut into sections make toothsome
roasting pieces. The bacon is cut into strips four or five
inches wide along the belly on either side. The thicker
parts of the sides are cut into pieces nearly square and
the fat edges cut off and made into lard. The shoulder
is cut smaller than indicated in diagram and rounded
off more like a ham, the trimmings being used for lard
and sausage. In trimming the hams be careful to avoid
cutting into and mutilating the natural membrane
covering the lean meat. A smoothly trimmed ham
looks better and keeps better than one that is rough
and haggled.

It is best as a rule to leave the cutting operation
until the carcass is cooled through and through, but
heavy hogs will cool faster split down the back, the
head cut off, and the leaf lard partly separated from
the sides.

Much pork can be sold in the winter season in the
form of the entire carcass, in a fresh state. When it
comes to salting down the meat great care must be
taken to have it thoroughly cooled. It must be entirely chilled through and through, but never put away
frozen. Do not spare the salt, as the meat will not
take it up in excess of a certain limit. Pickle should
never be poured into the swill barrel or where poultry
or any animal can get it to drink ; the surplus salt may
be spread over the asparagus bed.

In curing hams and shoulders be sure to fill in and

around the bones with salt and a little pepper, to keep off the flies.

For sugar-cured ham and bacon use six pounds of coarse or packers' salt, four ounces of saltpetre and from four to six pounds of brown sugar to each 100 pounds, and enough water to cover the meat when closely packed.   Boiling and skimming the pickle helps to clarify it.   Sprinkle a thin layer of salt on the bottom so that the meat will not come into direct contact with the wood.   Put the skin side down and be sure the whole contents are covered with the brine.  At the end of six weeks take up the meat and smoke it, using corn cobs or hard wood (green hickory is best) ; smoke until it is a light brown or tan color.   The pieces should be sewed up in muslin bags and painted with whitewash with a little ochre in it for coloring.   When prepared in this way bacon can be cooked without any freshening and it will keep sweet for a year or more.   It should be stored in a dry and cool place.

A few people prefer dry curing.   This requires the salt to be fine and the saltpetre to be pulverized. The ingredients are used in about the same proportion, mixed together.   After the meat is trimmed and cooled for thirty-six or forty-eight hours, place it skin side down on clean table or boards, and rub the mixture in thoroughly with the hands.   This must be repeated three or four times in two weeks, leaving a coating of the salt over the surface each time.   Bacon should be piled in layers four or five deep. After-treatment is the same as for that salted in pickle.

For making mess or clear pickled pork, use fifteen pounds of salt to 100 pounds of meat. Put a layer of salt in bottom of barrel and pack on edge, as shown in cut, sprinkling salt between each layer. Keep the meat covered with brine.

If we suppose pork in the carcass to be worth five cents a pound, there are many localities where it would pay well to work it up and sell it in various forms at higher figures. On a five-cent or seven-cent basis for the whole animal the bacon should bring ten cents, lard ten cents, head-cheese or scrapple ten cents, spare ribs ten cents, sausage twelve cents, shoulders twelve cents and hams fifteen cents. Of course the difference in price represents skill and labor, and it is well to convert skill and labor into money in this way whenever possible. Prices vary in different localities, of course, but it is everywhere true that the prepared articles are worth more per pound than the whole carcass.

When a ham is half fat, and the cost price is fifteen cents a pound, and no one will eat the fat, the ham really costs thirty cents a pound, and this makes it a very expensive meat. When hogs are fattened lean, or with only a reasonable percentage of fat, the meat is in every way superior and is in better demand in the market. The ideal hog, both as to cost and to please the present market, is therefore about as follows: 200 pounds, six or ten months old, and reasonably lean.

To try out or render lard so as to get the whitest

and firmest product it is necessary to remove all flesh, membrane and blood, and cut it into pieces an inch or less in size ; then heat a small quantity in kettle or boiler, and afterward add more and cook slowly. When the scraps, cracknels or cracklings are crisp, and a delicate shade of brown, strain the melted lard through a cloth into the vessel that is to receive it. Afterward the scraps may be pressed and an additional quantity of only slightly inferior lard obtained.   The fat from the internal organs and the small miscellaneous trimmings should be rendered separately and kept in separate vessels.   No water should be used during the operation ; in fact, the purpose of rendering the lard is partly to drive off all the water.

Excellent sausages may be made of the scraps and trimmings of fresh pork by using one-third fat and two-thirds lean meat, chopped finely, and seasoned with salt, pepper and sage.   The seasoning should be evenly distributed through the scraps before they are chopped.   As sage is offensive to some people it may be omitted.   A little red pepper is often added.   Tastes differ so much that a sausage receipt cannot be given that will suit all.   The meat grinders now in common use come handy for preparing sausage scraps.   Formerly the general custom was to soak the small intestines of the animal and cleanse by scraping thoroughly with a dull knife, and use these as cases for holding the sausage meat.   The meat being pressed into these by a "stuffer," they were twisted into links three or four inches long.   The practice still prevails, but much of the meat is now sold in rolls or lumps, wrapped in butter paper or corn husks.

Scrapple, a pork preparation well known in Pennsylvania, New Jersey and elsewhere, is made somewhat after the plan of souse and head-cheese, using portions of the head, ears, feet, liver, heart, and a part of the skin from which the fat has been cut for making lard. Not more than a fourth of the liver should be used. These scraps are put into sufficient water to admit of boiling a long time, at least until all the meat can easily be pulled apart by the fingers. It should then be dipped from the liquor and run through the cutter. When returned again to the boiler it should be thickened with finely ground corn-meal, seasoned to suit the taste, and thoroughly boiled, stirring vigorously all the while to prevent scorching. Some prefer to use one-fourth graham flour and three-fourths corn-meal for thickening. If the meat contains too much fat this may be skimmed off before adding the meal. Enough meal should be added to make a stiff mush when cold. Pour out while hot into square tins not over four inches deep. To prepare for the table, cut in slices and fry till brown.

Souse, made of scraps, especially including the feet, the lean meat of the head, etc., is prepared by boiling, and then flavoring with salt, pepper and vinegar, whole spices, etc. It becomes quite firm. It is browned in the oven for table use.

Head-cheese is much like souse, but the vinegar is replaced with sage, and the fat is pressed out through a strainer cloth. Like souse, it may be packed away in jars until needed for use. It is made ready for the table by browning in a frying pan.

To corn or pickle a small amount of pork for family

use prepare a liquid as follows : Take three gallons of water, four and one-half pounds of salt, one pound of brown sugar and one ounce of saltpetre. Boil for half an hour and remove scum. When cold pour it over the meat, and allow it to stand for several days.

There are many methods of pickling pork on a large scale, all depending on the preservative effects of salt, sugar, saltpetre, etc. The sugar may be replaced with molasses, and the meat may be afterward cured by drying or smoking. Every precaution must be taken to guard against insect attacks while the meat is in a fresh or partly cured condition. Every cook book gives a recipe for making pickle for pork, and there are as many recipes as there are cook books. I will give but one : To 100 pounds of meat use one pint of fine salt, four pounds of brown sugar and three ounces of saltpetre. Rub the meat thoroughly with this mixture, and allow it to lie for a day. Then pack in barrels, using additional salt freely. Drain the liquor from the bottom of the barrel and pour over the meat again. The meat should not rest upon the bottom of the barrel, but on a frame of some sort. After two or three weeks of this treatment the meat may be packed or smoked.

To prepare for smoking it is only necessary to wash off the brine, roll in bran (some people use sawdust), and hang in the smoke-house for four weeks. The house must not be permitted to become overheated. Good smoke is made with hard-wood chips and sawdust.

Hams may be packed in barrels or stored dry. The latter is the plan most in favor. One good way is to cover the hams with brown paper, then with coarse muslin, and then to give a coat of whitewash. A dark granary is an excellent storage place for hams.

I have before me a famous cook book which makes the sweeping assertion, at the beginning of a chapter, that pork is an unwholesome meat ; but admits that salt pork, bacon and ham are less objectionable than fresh pork. It must be because so many pigs are improperly fed and fattened that such statements are penned and printed, or else it is because human beings do not know how to adapt their food to their requirements. It is no doubt just as important to have a well-balanced ration in case of human food as in case of an animal's food, and if man does not suit his food to his habits of life he must suffer for it. Good pork, properly cooked and temperately eaten, is wholesome enough.

SCRAPPLE.

Ham half fat is too fat.

Fatten is a poor word. **Grow** is better.

Wholesome pork is digestible pork. Make it so.

Be merciful even toward a pig about to be killed.

Save the bladders. They make air-tight jar covers.

Save the bristles ; everything has some money value.

Head-cheese or scrapple is an excellent food when well made.

An unbalanced ration means wasted money, and perhaps inferior meat.

Extra quality bacon and hams rank among fancy groceries, and are ever in demand at top prices.

There are laws in some states against the sale of boar pork at current prices without explaining its character.

Wisely fed and fattened, cleanly, home-grown and home-cured pork is a deal better than the average market product. Depend on that.

CHAPTER XVI.

SMITHFIELD HAMS AND
DEERFOOT SAUSAGE.

Not how much but how good.—Martha.

 The celebrated Smithfield ham is so called from the little town of Smithfield, about thirty miles from Norfolk, Va. They have been cured there for nearly a century and their fame has spread at home and abroad. About 30,000 hams are now produced annually and are mostly sold direct to private families, some going to Europe. They are made from the half wild "razorback" pigs, which for a portion of the year run in the woods, thereby giving to the meat a gamy flavor not to be obtained in any other way. Many farmers are engaged in raising the hogs, but the killing and curing is in the hands of a few men in or near Smithfield. Many attempts have been made to "improve" the native breed of pigs by crossing, but in every case unsuccessfully, as it has resulted in a coarser grain to the meat, and the shape of the ham is not the same. The native "razor-backs" are thin-sided, deep-chested, with small flanks and long sloping hams. They are all colors. The sows run at large in the

woods and farrow about the middle of April. They
live on the nuts, roots and berries they find in the
woods. In the fall, after the corn, sweet potatoes and
peanuts are gathered, they are turned into the fields
and begin to pick up rapidly on what has been left.
They are then put in pens and fattened quickly by
giving them all the corn they will eat and pure water
to drink. They are kept clean and are killed when not
too fat nor too lean, weighing from 125 to 190 pounds.
They are carefully slaughtered, the hams being the
first consideration. The curing is as follows : The
hams are first placed in large trays of fine Liverpool
salt. Then the flesh side is sprinkled with crude saltpetre, using three or four pounds to 1000 pounds of
ham. The whole surface is then covered at once with
fine Liverpool salt. The hams are next placed in piles
not more than three feet high, and let stand for three
days. Each ham is then resalted with fine salt and
piled again, one day for each pound in each ham ; a ten-
pound ham thus stands for ten days. At the end of
this time they are washed with tepid water until the
hams are clean, and when nearly dry rubbed with fine
ground black pepper. They are then smoked slowly
and gradually for from thirty to forty days, using green
red oak or hickory. The hams are then repeppered to
guard against vermin and are bagged. The average
weight per ham is about ten pounds. These hams sell
for an average of twenty-two cents per pound at Smith-
field. The remaining parts of the hog are cured in the
same way and are largely consumed at home.

I believe there is no reason why with much the
same care and close attention to details, the same

results could not be achieved in other parts of the South where nearly similar conditions exist.

I visited not long ago the celebrated Deerfoot Farm at Southboro, Mass., whose hog products have an extended reputation and where they have no trouble in selling their hog products, and at an advanced price. Here some hundreds are slaughtered annually, and between the first of October and April the call for more is unremitting. Bacon, lard and sausage are the only outputs. Hams and shoulders find their way to market only in this form. Why? "Because our Deerfoot sausage brings more than any ham," said the gentlemanly foreman. "You see our links are a trifle longer and slimmer than other sausage. It is all made after one receipt, also every ingredient being proportioned by weight, so every lot tastes the same as that eaten before by our customers. We make it regularly, so that it goes to consumers new and fresh. One lot is not too salty nor another lacking salt. As a result, those who eat Deerfoot sausage or bacon find nothing else so satisfactory."

The sausage is wrapped neatly in parchment paper and tied, always the same, in one and two-pound packages. A two-pound package sells for fifty cents. The bacon is sliced and packed in pasteboard boxes, selling for sixty cents for the two-pound package. Upon each wrapper is printed the name of the farm and a few sentences that guarantee the quality, both as to flavor and healthfulness. Young pigs only are used, and if any are bought,—and some are bought,—the producers being engaged in advance to rear them for Deerfoot at an extra price, they must be reared within

one day's ride of the farm and strictly farm fed, not
fed upon refuse from cities nor at slaughter-houses
and breweries.

A GOOD BERKSHIRE—THE DEERFOOT KIND.

The delicious meat of the small-boned Berkshire
and Berkshire grade is used. The heads, feet and skin
all go to a cheaper market in bulk at low price. Deerfoot bacon is all dry-cured on the English plan and
hung for five days in the smoke of hickory wood cut
fresh every week from the forests of the farm. All lard
is rendered faithfully and never bleached artificially. It
is placed, not in a cellar nor in a cooler, but exposed to
the light in cases covered with wire fly net and aired
thoroughly while bleaching.

Chapter XVII

MARKET POINTS.

The hog is a machine for converting golden corn into golden coin.—John Tucker.

 In previous chapters I have introduced statements and facts which might have been reserved for this talk about markets, and shall here perhaps mention matters which might as well have been treated elsewhere. The truth is that every detail of the art of swine husbandry is so intimately associated with every other detail that it is quite impossible to consider each separately. Marketing, the final detail, may be called the sum total of the other details.

The profit in pigs depends very largely upon the age at which they are sold. The case was put forcibly by a trial in the West some time ago, when a cash experiment was made with fifty-four Duroc Jersey pigs. The average birthday of these pigs was April 15th. Their average weight, when weaned at two and a half months old, was forty-one pounds.

At eight months old their average live weight was 210 pounds, at a cost for feed and labor of $1.62 per 100 pounds.

At nine months old and three days their average weight was 247½ pounds, at a cost of $1.80 per 100 pounds.

At eleven months their average weight was 293 pounds, at a cost of $2 per 100 pounds.

Increase of weight was made at a continued increase of cost per pound. The pound cost in the three cases was 1.62 cents, 1.80 cents and two cents. The heaviest pigs, that is those longest fed, are therefore not the most profitable.

In the above case the growing pigs did not have much grass and clover in their pasture. A greater abundance of green food would perhaps have made a slight change in the figures, but the principle would have remained the same. The experiment is valuable because conducted on a large scale.

The day of profit in holding hogs for the block until eighteen months old is evidently past, never to return. If we had close commercial relations with the Esquimaux we might feed for fat alone, but our markets, both domestic and foreign, unquestionably demand leaner animals and a better quality of pork. It is possible to rear pigs so as to have seventy-five per cent. of lean meat in them. It depends principally on feed and exercise. If such pigs are demanded we must furnish them.

If pig meat is to take the place of hog meat, let it be so. With pigs, as with many other crops, there is much in knowing when to harvest. The animals must not be allowed to get too ripe. They must be sold when they will command the most money.

I think it is a good plan to have two lots of summer pigs ; one lot to go on the market when the animals weigh about 100 pounds, at a time when pork is often high, and the other lot to be sold later, say when about 200 pounds in weight.

There is certainly an increasing demand for small
hams, from pigs less than a year old, and it is obviously
cheaper to produce two fifteen-pound hams than one
thirty-pound ham. There is also an increasing demand
for good bacon.

In many places there is a call for choice lard, at
rates above market prices. Why not supply such lard?
Clover-fed hogs will yield the product, and if put up in

ESSEX SOW, TWO YEARS OLD.   ONE OF THE SMALLER BREEDS.

pails holding from five to twenty-five pounds, extra
prime lard will sell above current rates.

Many pork products, viewed from the standpoint
of a city consumer, are on the border line between
wholesome and unwholesome foods. Poorly prepared
or over-fat bacon, sausage, scrapple, head-cheese, etc.,
are voted down, while the same products from leaner
animals, if skilfully prepared for market, find ready
buyers. The less people dwell in the open air the less
fat they can digest, and it must not be forgotten that

city, town and village people consume and hence reg-
ulate the demand for a very large percentage of all the
pork sent to market by American farmers.

Ham is a standard food in this country, and is freely
eaten even by people who proclaim their inability to
digest pork. Choice hams are always in demand at
the highest prices paid for any portion of the hog's
carcass.

In the present condition of the market I shall allow
none of my hogs to get much above 200 pounds in
weight, except for special or temporary reasons.

We are getting along pretty well, as a people, in
learning how to produce things. We are not slow about
accepting the discoveries of science, and are ever ready
to harness nature's forces and put them to work in our
everyday affairs. But the Government is beginning to
recognize that another great national problem is before
us. "The rapid development of the agricultural re-
sources of the United States," says the U. S. Yearbook
for 1896, "has resulted in an annual production far in
excess of the consuming capacity of our population.
To such a degree has the surplus increased that its disposal is fast becoming a grave problem. The logical
solution lies in the extension of our markets beyond the
sea." The same volume elsewhere says that these
conditions justify the Department of Agriculture in
placing before American farmers as many facts and
figures relative to markets as it is possible to obtain.

My opinion is that while farmers need in no way
feel alarmed by the outlook, they should realize that
quality more than quantity will be the determining
factor in pork prices and profits during the next

decade. The farmer must suit the market, and fortunately the market calls for younger pork.

The Government tells us that "each year limited quantities of English bacon are shipped uninspected to New York and Boston grocers, who retail it at high figures to fastidious customers. It is considered a luxury at some American breakfast tables," etc. The same authority, our Secretary of Agriculture, says that American packers can only obtain and hold English and other European bacon markets by specially preparing their meats to suit the taste and demands of those markets. Smaller and leaner swine for bacon purposes are called for in nearly all foreign markets. And the meat must be mildly cured. But in Mexico and some of the South and Central American states the heaviest, fattest and thickest sides are required.

In the Yearbook for the following year, the newest one at this writing, the Secretary of Agriculture says : "Our bacon sells for less money in the English market than that of any other country. The reason for this is found in its over-fatness and saltness. * * * American hams are held in higher estimation than bacon and hold their own in competition with all other countries, so that in quantities shipped and in prices hams and pickled pork from the United States are equal to the same products from other countries."

Great Britain takes five-eighths of our hog product exports, and pays the United States over $50,000,000 per year for bacon, hams, fresh and pickled pork, and lard. Her trade is worth having, and the American pork packer may well try to please English tastes. Denmark is one of our keenest competitors in the line of bacon.

But though the foreign market is great, the home market for American pork is many fold greater, and it is in the home market that careful swine breeders must look for best returns, particularly with choice products, as I have already indicated.

Summing up the whole situation, from the cash standpoint, it is therefore evident that profits depend for the most part on economy of production. The quality of pork must be the best, yet there can be no food wasted in making it. Neither must any waste of the manure be permitted, for in many cases the real profits on the manure are fully as large as on the pork itself. Neither can farmers afford any losses through avoidable sickness among the hogs, for disease is excessively expensive.

Only the best methods will yield satisfactory results, and while I have my own choice as to breed of swine I think it is more a question of management than breed. It will no doubt sound presumptuous, but I cannot withhold the opinion that many of us are very wasteful of skim-milk and corn-meal in our ordinary feeding operations, by reason of our sluggishness in grasping the full significance of the idea of a properly balanced food ration. It is in reduced cost that we must look for increased profit.

On various occasions I have urged the selling of farm products in small packages in choice forms to particular people, and to learn the standing of the pig in really polite and select society I called the other day at a Chestnut street grocery store. The hog was

there, but in the form of bacon, ham and sausage ;
not as pork.

I found ham in tin boxes, packed in Chicago,
weighing eighteen ounces to the
box. The price was twenty-five
cents. The weight of the tin box
itself was six ounces, leaving twelve
ounces of ham. The consumer
therefore pays a little over two
cents an ounce for the ham, and
seems willing to do so. Not only
is the meat in an imperishable condition, but it is free
from bone.

For a tin box of imported German sausage
(Frankfurter) I paid thirty-five cents. This package
had a gross weight of twenty ounces. The tin can
weighed four ounces, leaving just a pound of sausage.
This sausage contained, I was told, two parts of pork
to one of veal. It was finely flavored, slightly smoked,
packed in skins, and in perfect condition when the
can was opened. This brand is regarded as a great
delicacy.

There are scores of other pork products on the
market, and I cannot pretend to enumerate them.
Those in tin may be had at any time of year, while
those packed in paper or pasteboard are of course
limited to the cooler months.

I think there is room for individual enterprise in
the manufacture and preparation of special brands of
ham, shoulder, bacon, lard and sausage for the fancy
retail trade of every large city. In some respects
the great packing houses have the advantage over

individual operators, but it must not be forgotten that the man who fattens his own hogs (if he does the work properly) has an opportunity to make better pork than the average on the market.

So I repeat that I am greatly in favor of individual effort in the production of really choice food products. There is always a premium on such products. There are many buyers whose first question refers to quality rather than to quantity or price, and these buyers constitute what is known as the fancy trade. Cannot my readers get a share of that trade?

COINS.

There is a decreasing demand for overfat hogs.

If light-weight hogs pay best in money, why do you raise heavy porkers?

Put the hogs on the platform scales occasionally. You will learn something.

Watch the markets. Notice the ever-increasing demand for good goods in small parcels.

The pork market is often temporarily depressed, but it will never fail entirely. Pork is one of the standard foods.

England is the greatest foreign buyer of American hogs. It is therefore worth while to recognize English ideas about bacon.

Profit comes not in how little we can keep the pig on but how much we can get him to eat of a balanced ration.

If figures are to be believed, it costs all the way from one and one-half cents to seven cents or more to produce a pound of pork.

Market some of the young pigs for roasters when they will dress twelve to fifteen pounds. City people will be glad to get them.

The younger the animal the more thoroughly it digests its food, therefore mature pigs early. The six months 200 pound pig costs one-half the 200 pound eighteen months pig.

Chapter XVIII.

THE POOR MAN'S PIG.

A pig may increase a pound in weight every day, and a pound of pork per day is enough for a family.—Tim.

The man who keeps one, two or three pigs usually has a different problem to solve from the farmer with a larger number of animals, for in one case the pigs are fed upon swill and refuse, while in the other case they are pastured or fed upon products of the farm which might otherwise be sold. The man with one pig saves wastage.

The one-pig man must first consider the food supply. Very often the refuse from the family table, especially if one or two cows are kept, will be nearly sufficient. In such cases a so-called swill barrel is made the receptacle for every sort of refuse food material, often including the dish water.

Swill is a proper pig food, if not allowed to become stale and foul, especially if balanced as to contents with middlings or corn-meal, as the case may require. If the swill be mostly milk, a little corn-meal should be used regularly. If it be mostly water, food scraps, fruits and vegetables, there should be some middlings added to it, to make it more nutritious during the growing period of the pig.

As to the use of dish water, people must do as they please. I prefer to use it for fertilizing purposes, on sod, as its soapy constituents cannot have much real food value. Fresh clean water is better for the hogs. If the dishes are carefully cleaned before being washed the dish water will contain but small traces of food.

The pig pen for one should be movable, so as to be easily and thoroughly cleaned. It should not be a rat harbor. It should afford warm and dry shelter to the pig. And, finally, it should be built with the idea of saving all the manure that is produced.

It is not necessary to construct a costly building. Posts may be lightly planted, with a view to the future shifting of the house and the pen, and the ground given to the pig one year may next year be plowed or spaded up and planted with vegetables.

No elaborate care of the manure is necessary, but if it be daily or frequently scraped up and put under the temporary shelter afforded by a few boards it will be found to accumulate very rapidly ; and being mixed somewhat with soil it will be in excellent order for preservation. In feeding try a little bone meal, in a box separate from the feeding-trough. If the bone meal fails to satisfy the animal's craving for bone-making food (if such be the object in rooting) try wheat middlings. Possibly charcoal and salt will quiet the pig. Try these several things before using the half-cruel ring.

The shed need not be expensive. The floor should lift out bodily, which will make it easy to hunt the rats and to move the building. The house itself may easily be made in sections, to be taken apart at pleasure. Three sides of the house should be tight and the other side open, but so arranged as to be closed or partly closed during cool or cold weather. The trough and swill dump may be temporarily secured by stout stakes, and everything made as satisfactory as though the quarters were intended to be permanent. This is the way to have clean, healthy pigs, and it will pay every time.

A little bedding is good, even in mild weather, for it is only in midsummer that our nights are really warm, and a shivering pig is not a growing pig. The animal must be comfortable as well as clean in order to do its best.

It costs but little to indulge in the luxury of two swill barrels, and to keep one of them *always empty and sweet.* In practice I never could get one barrel emptied and scalded at regular intervals without wasting food, but now it is a simple matter to keep things reasonably clean. Besides, it is a comfort to always know that there are no germs of any kind more than a week old in the swill barrel. Disease germs cannot withstand frequent changes. It is only in neglected places that they flourish. Sunshine is hostile to all disease germs.

Shade, pure water and green food are essentials to the most rapid growth of penned pigs. Each of these items counts for much, and yet each is often neglected. Some people never give their pigs pure water to drink, compelling them to rely wholly upon the not too savory swill ; and the mistake of withholding green food is almost as common. The argument that pigs live and grow under such treatment is no proof that they would not do better under the wiser way suggested.

I think that much milk is lost simply in quenching thirst; thirst that could as well be appeased with water. But when people have the milk to dispose of as a by-product, and do not know how else to use it, I suppose they will continue to pour it into the trough in excess of the digestive powers of the pig. It is quite important, however, that the owner of even a few pigs should be made aware of the real digestive require-

ments of the animals under his care.

This information is briefly tabulated in the chapter on feeding, and here I will merely say that a 100-pound hog requires only 3.4 pounds of organic matter (water

OUT FOR A LITTLE WAYSIDE PASTURE.

free) per day. This amount of organic matter would be contained in twelve pounds of skim-milk and three pounds of corn-meal. Such a ration would be suited to a pig somewhat above 100 pounds in weight.

Swill made of house scraps is probably as rich in food elements, on the average, as a mixture of skim-milk and corn-meal in the proportion just suggested.

As to the merits of salt, charcoal, bone meal, dried blood, offal meat, etc., for pig feeds, there are different opinions and practices. Swill-fed animals which

receive the broken bits of human food get a good deal
of salt in that way.   Mixed salt and charcoal is sometimes a useful condiment or appetizer, especially where
the hog's ration has not been perfectly balanced, and
where by reason of restricted quarters it cannot
search for food adapted to its cravings.   Ground bone
and dried blood are sometimes of great use as side
dishes (not in the trough) to afford needed nitrogen
and phosphoric acid ; in other words, to supply muscle and bone-forming materials where the diet has
been too largely of corn or other carbonaceous food.
Offal meat of any kind has no right place in the pig
pen, and is distinctly liable to cause disease.   Such
meat, including entrails of butchered animals, dead
chickens and rats, should always be buried or com-
posted.   The soil is the true place for them, for they
contain much fertilizing value.

When pigs which are kept alone, under good
treatment, fail to make rapid growth it is because of
improper nutrition, and the swill should be supplemented either with wheat middlings, or whatever
nitrogenous food may be cheapest. A very small
amount of cottonseed meal, a few ounces only per
week, may be given to a pig which has not sufficient
nitrogenous food to make rapid growth ; but wheat
middlings would perhaps be better and safer.

When it comes to finishing the home-raised
porker, and making ready for the block, we have
nothing in America superior to corn ; and corn should
be fed freely for several weeks before killing. Indeed,
if the swill diet can be wholly replaced for a month before slaughtering time by a diet of corn, it will do

much toward rounding the hog out, increasing the weight, and adding to the firmness of the flesh.

The taste and skill of the owner must determine how best to put on the finishing touches—whether to seek for additional fat by heavy feeding of corn or only to seek for a little more plumpness to a hog already in good condition. There is a general belief that corn has a decided influence in improving and sweetening the pork, and it is very common to finish the feeding in the way suggested, though many feeders give some swill even to the last.

Butchering in one-hog establishments is sometimes done in the old-fashioned way, with every detail performed at home ; but now-a-days, in the Eastern states, it is perhaps more common to send the animal to a slaughter-house and pay a dollar for having it killed and dressed.

———

SCRAPS WORTH SAVING.

A dry shed and a dry bed.

Feed only what will be eaten up clean.

Water is the cheapest element of pork.

Leaves make good bedding, but straw is better.

A squealing pig is cold, hungry or uncomfortable.

Spare no trouble to start with good juveniles. It is half the battle.

Remember: $12 worth of manure for each pig per year, if not wasted.

If a hog's manure is worth $12 per year, as estimated by U. S. bulletins, it amounts to just a dollar a month. Do you get such a dividend ?

Lice multiply in muggy weather, amid unclean surroundings. Receive them with lard scented with kerosene or tobacco. Then clean the pen carefully and use fresh bedding.

Chapter XIX.

THE MANURE PILE.

Half the hog manure is lost and the other half is too often neglected.—John Tucker.

YORKSHIRE BOAR.

More than once in the previous pages have I referred to the fact that a recent estimate places a value of $12 on the yearly manure product of a hog. The average of each horse is estimated at $27, for each head of cattle $19, for each hog $12, and for each sheep $2.

The manure value seems almost incredibly large, for what farmer with ten hogs counts on getting $120 worth of manure from them each year? Indeed, what farmer clears $12 per head on his pork, above the cost of feeding? We often hear of profitable pork-fattening operations, but much less is said about the hog manure or its cash value. Perhaps there is no better way of showing the value of the pig-pen product than by comparing it with the product of the cow stable.

	Water per cent.	Nitrogen per cent.	Phosphoric Acid per cent.	Potash per cent.	Value per Ton
Cow manure . . .	75.25	.426	.290	.440	$2.02
Pig manure . . .	74.13	.840	.390	.320	3.29

Pig manure, however, is variable in composition,

due to the mixed nature of the food supplied to this animal, but is generally rich, although containing a high percentage of water. It generates little heat in decomposing.

The urine is valuable, but not so valuable as the manure itself. This is exceptional, as with other domestic animals the reverse is true. Still, hog urine should be carefully saved along with the other.

The argument will of course be advanced by busy people that it does not pay in money to expend so much time and labor in the hog yard as would be involved in the daily collection of all the droppings and in saving the urine.

The answer is that such collection should be auto-

matic, as much as possible. Manure dropped in large yards or pastures is not lost, as it finds its way directly to the soil, and will exercise a wholesome influence as a fertilizer. The pig pen and buildings should be

WESTERN HERD OF JERSEY REDS.   shifted from time to time,
so that the highly enriched location occupied by the pigs will in turn come under the plow.

Drainage should be carefully attended to, so that all leachings will flow over sod, or over garden or field soil.   Little or nothing will be lost if these precautions are observed.

The bulk of the manure, including all which is made in the shed and all which can be easily scooped up in the yard or pen, containing more or less litter, should be most carefully accumulated, either by storage under a shed, or by building it into a compact heap, or by frequent removal to field or garden, to be spread whenever convenient.

Hog manure is a heavy product and quite hard to handle on account of its weight, and hence the work of caring for it is expensive. I know that labor can be ill spared on the farm for anything except the necessaries, but I come back to the point of beginning and urge that it is necessary to take care of a product worth $12 per year per animal.

Do anything with hog manure except waste it.

The common practice is to allow the manure to accumulate under the hogs, layer by layer, and to haul it out only once or twice a year. This practice is not a bad one under some circumstances. If the pig pen is surrounded by a solid stone wall, so that no water except rain can enter, and no leachings can escape, and if there is plenty of litter to be worked up, it may be a good plan to allow the manure to thus accumulate. But I always fear the contamination of some near-by well of drinking water, to say nothing of the injury sustained by pigs compelled to perpetually breathe the products of fermentation and to lie down upon couches that are always mouldy and often wet. Such manure beds, with pigs upon them, often occupy the basements of barns.

It is time to regard the pig as a clean rather than as an unclean animal ; and I think the markets will compel this change of treatment, for cleanliness is directly in the line of choice pork products.

It is some labor to make a stack of hog manure, to scatter plaster upon it, and to put it into neat shape. But it probably pays well. I have personally done it with satisfactory results; certainly with the result of having more and better manure than when the operation was forgotten or neglected.

Taking the fertilizing value of equal weights of manure in its natural condition, farm animals probably stand in the following order: Poultry, sheep, pigs, horses, cows.

———

GAINS.

Be jealous guardians of all manure made on the farm.

Hogs drop most manure quite near their feeding places.

The hog loves a bath, but what benefit is a bath in liquid mud?

Hog pens should not be on steep places; too much manure is lost.

Clean the pen twice a week, and be surprised by the size of the accumulated heap.

The dung of hogs should never be in pellets, as such a condition indicates constipation.

Chapter XX.

HOG CHOLERA.

In hog cholera an ounce of prevention is worth ten pounds of cure.—Tim.

THE BARS THAT KEEP HOG CHOLERA OUT.

More has been spoken and written on the subject of hog cholera than upon any other one subject connected with hogs. It has ever been a fruitful source for discussion at farmers' institutes and an endless theme on which to write. The Government has appropriated large sums of money and has employed learned men who have labored with seeming diligence for years, and yet after all these years of waiting and all this expenditure of money we are forced to admit, whether humiliating or not, that we know but very little that is of practical benefit about the whole matter.

But two things are absolutely known about the disease. One is that it sweeps unrestrained over vast areas of country, leaving death and destruction in its wake ; and the other is that hogs which contract the disease usually die.

I shall not attempt to deal with this subject in a scientific way, but shall deal with it rather from a practical standpoint. Much that I know, in fact most all that I do know, has been learned in the school of bitter experience, and the lessons were sometimes very costly.

Before entering further into this subject I desire to notice what to me is the most hopeful promise held out by any of the investigators. In this, as in all other great searches after truth, some men have stood far in advance of their fellows. The only ray of hope I see held out to the swine raiser comes in the promise of immunity through prevention. Some of the investigators have contended that the animals could be rendered cholera proof by inoculation with a virus containing the germs of the disease especially prepared for the purpose. They have not as yet been able to render this practical, for the reason that the virus would sometimes produce the genuine hog cholera and kill the hogs operated on. The unsolved problem seems to be to cultivate the germs in a form mild enough.

That they will in the end be successful I have not the slightest doubt. How long we may have to wait before they will attain success I cannot say, but that they will succeed in the end I know from what is generally conceded in regard to the disease, and that is that a hog once having the disease will never take it a second time. This being the case, all that remains to be done is to cultivate a virus sufficiently mild as not to endanger the life of the animal and of sufficient strength to produce the disease in a light form.

Another and somewhat more recent means of preventing the disease is the serum or antitoxin cure. It consists in introducing into the system of the animal a serum which enables the body to more successfully combat the disease. The Government officials seem to be highly pleased with the results so far and seem to believe that relief from the dread disease is likely to come through this means. The serum produced last year, wherever used in cholera-infected herds, saved over eighty per cent. of the animals. It is easily applied, and its good effects in sick hogs are seen almost immediately.

That hog cholera will ever be successfully treated with medicine I doubt, but that it will be prevented in time I firmly believe. The whole trend of investigation seems to be in that direction, and I feel sure success awaits their efforts. Till then I suppose we must make use of the best means at our disposal to combat the disease.

Page after page has been written as a means of telling hog cholera, but much of it is difficult of comprehension to the average reader. If you have never had it in your herd you are to be congratulated on your good fortune ; and if you ever do, when you are done with it you may not have as many hogs as you did before, but rest assured of one thing, and that is you will know hog cholera when you see it again. As a rule hogs do not look well for weeks before an attack. At other times it will come like a bolt of lightning from a clear blue sky. The first thing noticeable is a loss of appetite ; the hair will look harsh and dry ; sometimes a slight cough will be noticeable, at other times not. The disease is sometimes of slow development, at other times quite rapid. Instead of the sprightly, rapid movement so characteristic of the young and growing hog, he moves slowly and indifferently; he looks gaunt and tired ; his back is arched, and he moves his hind legs with a dragging motion; his temperature will most likely be high, probably from 104 to 108—the normal temperature of the hog is from 100 to 102. His bowels may be costive or the discharges may be thin and watery in substance, but usually black or dark in color, emitting an offensive odor peculiar to the disease.

The disease may be of a lingering character and the animals linger for weeks, or they may die in three or four days. Usually the lingering type is less fatal than the more rapid forms of the disease. Hogs which discharge freely in the first stages of the disease are more likely to recover than when the bowels remain corstipated. Dark blue spots will often appear under the skin. The bowels will be more or less inflamed inside ; in the small intestines and sometimes in the stomach will be found ulcers ; this, however, is not common in the first stages of the disease. The bladder will most likely be full of a dark thick substance, showing that the kidneys, and in fact the whole internal organism, are affected.

If I were to say what I thought was the best thing that could possibly be done when cholera appears in a herd, I would unhesitatingly say, take the well hogs to clean new quarters where no hogs have been for years. Then if more of them take sick move them again, and it is my belief based on actual experience that more can be accomplished in this way than by the use of all the medicine in the country. For various reasons it is not always possible to move hogs, and in that case treatment may be resorted to, sometimes with fairly good results. The treatment should consist in separating the well from the sick hogs, and in dividing the sick hogs according to age and size and severity of the attack. I do not think that more than four or five hogs should be in the same pen, and fewer would be still better. Feed but little, and let that be food which is easily digested. Use air-slacked lime and crude carbolic acid freely as a disinfectant. Use it both on the hogs and on the ground, in the sleeping places, on the fences and in the drinking vessels. As much depends on a thorough use of disinfectants as upon any other thing. If the bowels are constipated give something to move them. If too loose give something to check them. In short, use good common horse sense (so to speak) and you will usually succeed very well. I have found nothing better than salts or oil to move the bowels, and I have tried nothing with better results to check them than a few drops of crystal carbolic acid. I know of no food better, if indeed as good, for sick hogs than ship stuff, or middlings as it is sometimes called ; it seems to digest easily and is soothing to the bowels.

If the weather is wet and cold keep the hogs dry and warm. In wet weather (if not too warm) keep the hogs in a floored pen, or at least in a pen where no water will lie in sinks or holes, as dirty water is one of the worst things a sick hog can possibly have. If the weather is warm, shelter the hog from heat. In other words, make him as comfortable as possible.

Let it be borne constantly in mind that much depends on good nursing. It would seem natural and reasonable that an animal afflicted as he is would do best if allowed plenty of fresh water to drink, but actual experience demonstrates that a greater number recover when the supply of water is limited than when it is not. I do not pretend to say why it is so, but experience has taught me that it is.

Hogs that are very sick should be kept by themselves, as others seem to disturb them, and often their recovery depends on being perfectly still at the critical period of the disease. I have never been successful in drenching hogs ; I have sometimes done it, and sometimes they recovered and at other times they did not, but even when they did recover there was nothing to prove that the drenching had anything to do with it. As a rule hogs that are too sick to eat die. All hogs that die of cholera, or of any other disease for the matter of that, should be burned and not buried, as abundant evidence can be produced to prove that the carcasses of hogs dying of cholera have been the cause of an outbreak years afterward. Hence, I say by all means burn all dead hogs as the only absolutely safe way of disposing of them. The burning operation is very simple. Lay the bodies across two logs, sticks or pieces of iron that will keep them up off the ground so that the fire can get under them, and the grease from their own bodies will usually do the work, with a little wood or corn cobs added occasionally.

How can we guard against the disease so as to prevent it is a question easily asked but not so easily answered. Men with medicine to sell will tell you they can, but my belief, based on bitter experience, is that they cannot.

Experience teaches that the disease more commonly appears in large herds than in small ones. The moral of this, then, is easily understood. Do not keep hogs in large droves. I do not believe that over twenty-five or thirty hogs at most should long remain together, and half the number would be infinitely better and safer in every way. Hogs of different sizes and ages should not be kept together, excepting of course sows and suckling pigs. Hogs should not be kept on the same ground from year to year if it can possibly be avoided. Plow up the lots and pens and cultivate them for a year or two ; it will greatly assist in keeping your lots free from the germ. The disease is much more prevalent in the summer and fall months than in other seasons of

TOO MANY CORN COBS HERE.

the year. Then as far as is possible reduce the number of hogs on the farm at this season of the year.

If your neighbor's hogs have the disease, stay away from his pens and be sure he stays away from yours. Shoot a crow, a buzzard, or a stray dog that comes on your place as unhesitatingly as you would kill a mad dog. This trio in my opinion does more to scatter the disease than all the other causes combined. If your hogs are fit or any way near fit to go to market when the disease makes its appearance in the neighborhood, sell them without delay. "A bird in hand is worth two in a bush." If your hogs have cholera this year, don't get discouraged and quit, but try it again, on fresh ground.

If your brood sows have passed through the cholera, keep them ; they are valuable. They will never again have the disease, and their pigs are not nearly so apt to contract it as pigs from sows that have not had the disease. Look out for streams which come down from some neighbor above you. This has been found a frequent cause of cholera outbreaks. The germs of hog cholera possess great vitality, and will live in the soil, in moist matter and especially in water, for months.

If you feed corn, rake the cobs together often and burn them; pour water on the coals and then put salt on the charcoal thus made and you have an excellent preventive for diseases, with little or no cost. Keep your hogs, excepting brood sows, ready for market. It may come handy some day. Strong, vigorous hogs are less liable to contract the disease than hogs of less strength and vigor. Then breed and feed for both these things. Eternal vigilance in hog breeding, as in other kinds of business, is the price of success.

Here is a formula for the treatment of hog cholera that is probably as good as any, which is not saying much. It is suggested by the Department of Agriculture:

Sulphur	1 pound.
Wood charcoal	1 pound.
Sodium chloride	2 pounds.
Sodium bicarbonate	2 pounds.
Sodium hyposulphite	2 pounds.
Sodium sulphate	1 pound.
Antimony sulphide	1 pound.

Thoroughly mix and give a large tablespoonful to each 200-pound hog, once a day. If the animal does not eat, add the medicine to a little water, thoroughly shake and give from a bottle by the mouth. If the animal will eat, mix the medicine with sloppy food. The same remedy is recommended as a preventive to those animals that do not as yet show signs of disease.

If you have had cholera on your place, and you have small, inexpensive pens, burn them at once. In a piggery, burn all the litter and loose inexpensive parts; renew the floor, if possible, and disinfect the remainder by washing it with hot water and washing soda. After washing, apply with a whitewash brush, or better yet a spray pump, a solution of one part of carbolic acid to fifty parts of water. Then thoroughly whitewash. Treat the fences in the same way. Earth floors should be removed to a depth of at least six inches and the ground sprinkled with chloride of lime and a few days later a good coating of air-slacked lime. Don't put pigs in the quarters for at least six months, and, if possible, have them vacant over the first winter.

An Ohio breeder of large experience, in the Miami valley, where hog cholera first appeared in 1856 and has recurred at frequent intervals, holds that drugs, virus and antitoxin have all been fairly tried sundry times by him and his neighbors. He believes that prevention will do more to hold in check the plague than drugs and hypodermic infusions. The most important help to prevent spread of disease is not to allow the hog farm to become infected with the excrement of diseased hogs. This can be done by quarantining the herd in a field, that is to be put under cultivation the following year. This quarantine must be established as soon as the first pig is taken sick. If the disease is in the neighborhood, carefully watch for first symptoms of disorder. Do not wait until several are sick and scouring, for this excrement is loaded with germs of disease, and these germs may retain vitality many months when covered in the corners of pens, or filth of yards, or about an old straw-stack ; but when exposed to sunlight or dryness they lose vitality in a few days, and under some very drying sunlight conditions in a few hours. Carefully observing these facts, he has in forty years been clear of hog cholera the year following an attack, and on until the disease has become epidemic in his neighborhood. After the herd has been placed in quarantine away from the permanent hog houses, lots and feeding floors, he kills and burns, or buries five feet deep, each animal as soon as it shows distinct symptoms of disease. They are burned or buried beside the quarantine, and in the field to be cultivated the following year. It requires nerve to kill breeding stock of great value, but they are as liable to spread and entail disease as any other, when once attacked.

If, by any means, we can prevent spread of germs, by so much do we hold the disease in check. A farm, with its feed lots and pens and shelters infected by the excrement of the diseased, becomes as deadly a centre as the public stock-yards and filthy stock cars on the railroads, and these are so thoroughly infected that we can never safely take stock hogs from these to our farms. This is not theory, but well proven fact.

The large hog houses and adjoining lots, once infected by sick hogs, are so difficult to disinfect and make safe, that I have abandoned the use of mine except for feeding off the fatter hogs in the fall or early summer. As soon, then, as the drove is sold, the house and lots and feeding floor are cleaned and disinfected, and the houses and fences whitewashed and left open to sunlight and drying winds ; the brood sows and young pigs are not allowed there. By having a few portable pig houses for sows to farrow in, and kept there until the pigs are weaned, the houses can be taken down and whitewashed inside and out and exposed to sunshine and winds to destroy all germs. The lots in which the little houses were set are put under cultivation, and the houses set up again on new ground for next farrowing season. When any new breeding stock is brought to the farm, it is placed in a lot with one of these houses remote from other hogs, for at least three weeks.

The theory of removing well hogs from sick ones is defective, and fails because there are in fact no well hogs in a herd where even one sick one has been over night. The only safe way is to treat all as diseased, and confine them to one field or lot that is not to be used by other stock until after a year or more of cultivation. By such care we can keep our farms healthful places, and go on growing and feeding hogs each year, and not be guilty of making our farm a new centre of disease. If, in addition to such preventive measures, we could keep from our farms buzzards, birds, fowls, crows, opossums, dogs and men, that can readily carry the deadly germs of disease, then we should not need any of the many nostrums that are adversised as sure cures.

It is hard to get farmers to understand the nature of the deadly germs, and to cease harboring and multiplying them in the old feed lots and pigs pens, all so handy to the barn, corn crib and kitchen that they cannot be purified by fire as they need to be.

When hog cholera once appears on a farm, the first thought should be not what cure can we get, but how can we keep the plague from becoming perma- nent, as it has become on many a farm, until the owner has ceased to grow hogs.

A FEW CHOLERA DONT'S.

Don't let your hog drink dirty, filthy water.

Don't castrate pigs when cholera is in the neighborhood.

Don't bring home cholera from the fairs and stock shows.

Don't wait until your hogs are all dead before doing something.

Don't forget to disinfect all quarters where sick hogs have been.

Don't throw dead animals in a creek or river. Burn them every time.

Don't put pigs in a field where there has been cholera for at least a year.

Don't drag a dead animal over the ground. Carry it on a plank or in a box and burn all.

Don't keep your hogs in a field along a railroad if you can help it. Railroads often spread the disease.

Don't overcrowd. It is responsible for many troubles and multiplies directly the chances of all contagious diseases.

Don't forget to be considerate of your neighbors. If you have cholera, put up a sign, "*Hog Cholera, Keep Out,*" and insist upon it.

Don't immediately introduce new animals into your herd. Put them by themselves awhile, at least three weeks, until you are sure they have no cholera about them.

Don't fail to have plenty of charcoal around where the hogs can get at it. It acts as a condiment and preventive. An excellent plan is to use up the corn cobs in this way. Gather them into a pile, and when they are thoroughly ablaze, put out the fire by throw- ing water or earth over the pile.

Chapter XXI

OTHER PIG AILMENTS.

It is hard to doctor a sick hog. Better never let them get sick, by giving range, pasturage and a chance to be natural and keep clean.—John Tucker.

 Pig ailments are numerous; I shall speak only of some of the most common. I do not believe the great majority of the readers of this book care to know the scientific names of the different diseases. I believe that they will be more interested in knowing how to tell and how to treat them than to be able to call them by their scientific names, hence I shall leave the technical names out of the discussion.

It is always best to give medicines mixed with food or drink where possible. If the animal refuses food or drink and it is necessary to administer drugs, it may be done by placing a stout chain (an ordinary harness breast chain does very well) within the mouth and well back between the jaws, which are thus kept from crushing the bottle. Two or three men are necessary for the undertaking, one or two to hold the chain and one to pour the medicine. The head should be well elevated, which places the pig on his haunches. Do not pour the medicine fast enough to strangle the animal.

Hogs will not do well when the skin is covered with filth. Bad air will bring on coughs; all corn for food, fever ; a wet bed, rheumatism ; and a big bunch together will breed disease. With a clean skin, good air, a variety of food, a dry bed and a few together, and lots of out-of-doors, they will do well.

When at pasture they find many roots, nuts and pebbles, besides being continually active, which does more than food for their hearty health, rapid and easy digestion and speedy, profitable growth.

I hope that American farmers who raise hogs on such foods as grass, clover, grain and milk will lose no opportunity to condemn the feeding of pigs upon slaughter-house refuse and such disgusting, offensive and disease-breeding material. Hogs fed on the offal of animals are only too liable to be infested with trichinæ ; and the whole idea of giving such stuff to hogs is a wrong one, tending only to bring the use of pork for human food into disrepute. Slaughter-house waste should be converted into fertilizer, of course ; not given to pigs and rats, nor allowed to go to decay. Some of our neighborhood slaughter-houses are discreditable ; yet thither not a few hogs are taken for butchering. The Government is beginning to point out some of these evils through its meat inspectors.

THUMPS.—This disease is quite common (especially in the early spring) and is exceedingly hard to handle when once contracted. More can be done to prevent than to cure. You visit the sow and litter in the morning to give them their accustomed feed, and you notice that one of the fattest and plumpest ones does not leave his bed as do the others. You enter the sleeping room and compel him to come out, which he does somewhat reluctantly, and you will notice that his sides move with a peculiar jerking motion, and if allowed he will soon return to his bed. Rest assured he has thumps, and nine chances to one he will die. It is caused by fatty accumulations about the breast, which interfere with its action, and the lungs work hard —pump for dear life to keep up the heart's action—to send the blood through the body. The pig is faint because of feeble circulation, and he is cold, and soon dies from exhaustion or weakness. He has no strength to suck or move. Poor little thing!

To prevent thumps, get over into the pen several times a day and hustle the little pigs about the pen ; also stint the sow so that she will give less milk. Pigs when they stir about, and when they are thin in flesh, rarely have thumps.

I have sometimes succeeded by shutting them out in the sunshine for an hour or two each day, but usually they die. Thumps rarely occurs among pigs farrowed after the weather is fine, but does quite frequently occur among pigs farrowed in the early spring. If the weather is cold and stormy and the sow and litter keep their bed much, then be on the lookout for thumps. Guard against it by compelling both sow and litter to exercise in the open air.

CANKEROUS SORE MOUTH is a disease which is quite common and which if not promptly taken in hand is often quite fatal. When pigs are from a few days to two weeks old, you may notice a slight swelling of the lips or a sniffling in the nose. An examination will show a whitish spongy growth on the sides of the mouth just inside the lips or around the teeth. This is cankerous sore mouth, and if not taken promptly in hand will result in the death of the entire litter, and will sometimes spread to other litters.

Some claim the disease is caused by damp and filthy beds, others say it comes from a diseased condition of the sow, and still others claim it is caused by the little pigs fighting over the teats and wounding each other with their sharp teeth, and stoutly aver that if the teeth are promptly removed no case of sore mouth will ever occur. I am inclined to believe there is some truth in each of these claims. I do not believe that the wounds made by fighting will alone produce the disease, but it is quite reasonable to conclude that the wound furnishes a place for the germ to begin its work.

Hold the pig firmly and with a knife or some cutting instrument remove all the spongy foreign growth, and be sure you get it all even though the pig may squeal and the wound bleed ; your success in treating the disease will depend largely on the thoroughness with which you remove this foreign growth. After removing the fungous growth apply an ointment made of glycerine and carbolic acid in about the proportion of one part of the acid to from five to eight parts glycerine. Repeat this each day for three or four days and the disease will usually yield. You may discover in a day or two after commencing treatment that you did not succeed in removing all the cankerous growth at first, and if so, repeat the cutting operation till you do remove it all.

Another treatment which I have heard recommended but which I have never tried, is to catch the diseased pig and dip his nose and mouth up to his eyes in chlora naptholeum without diluting it. This is certainly easily done and is highly commended by the person suggesting it.

Is the disease contagious? I do not believe it is in the usually accepted meaning of the term. I have often had one litter affected and other litters in an adjoining pen show no signs of the disease. Hence I have concluded that while it is possible for one pig to communicate the disease to another of the same litter, I think it quite improbable that it will be communicated to one of another litter. So firmly do I believe this that when I find one litter affected I give myself no uneasiness about other litters. Pure peroxide of hydrogen, applied with sponge or syringe after removal of the fungous growth, is very good treatment.

BLIND STAGGERS, INDIGESTION, SICK STOMACH, FOUNDER.— Causes, over-feeding, especially common with new corn; sour or decayed food. Sudden warm sultry weather predisposes in highly fed hogs. Insufficient exercise is also a predisposing cause.

Symptoms.—Loss of appetite, bowels constipated, or maybe diarrhœa. In some severe cases blind staggers and great paleness of mouth and nose, coldness of surface of body; abdomen may be distended and drum-like from contained gases.

Treatment.—Remove sick animals, provide clean, dry, well ventilated quarters, with chance for exercise, and fresh earth and water. If animal will eat, give light feed. Give charcoal in lump form, also mix soda bicarbonate in food at rate of two tablespoonfuls per day to each half-grown animal. It is rarely necessary to drench with medicine. If recovery begins, use care not to again feed too much.

MILK FEVER occurs in sows immediately after farrowing or within the first few days afterwards. The symptoms are loss of milk, swollen, hard condition of the milk glands, which are more or less painful on pressure. Sow may not allow the pigs to suck ; she may lie flat on her belly or stand up, and in extreme cases the sow has spells of delirium, in which she may destroy her young.

Cause.—Injudicious feeding, overfeeding on milk-producing foods. Do not feed sow quite full rations for few days just before and after farrowing.

Treatment.—Give sow plenty of cool clean water ; bathe the swollen glands for half hour at a time with water as warm as she will bear, dry thoroughly with soft cloth and give good dry pen. If bowels seem constipated give the sow internally one-half pint pure linseed oil. (Never use the boiled linseed oil used by painters ; it is poisonous.) If the sow starts killing her young, or has no milk for them, it is best to take most of them, or all, away from her and feed by hand with spoon or ordinary rubber nipple and bottle. For this use one part boiled water and three parts cow's milk. The pigs may be returned to the sow if her milk returns.

SCOURS among pigs is another common and very troublesome though not dangerous disease. This disease is not confined to any particular season, but is more common in the wet, damp weather of April and early May than in other seasons of the year.

As in thumps, remove the cause. This disease is almost in-variably caused by some improper food eaten by the sow. A sour swill barrel is often the cause. But to be doubly sure that I make no mistake in removing the cause I stop all feed and give nothing but water to drink and possibly a little dry corn to eat for three or four days, and I seldom fail to bring about a speedy cure. It should be borne in mind that pigs once affected will be more liable to a recurrence of the disease than those never affected, and greater care should be used with them for some weeks till they fully recover.

CONSTIPATION.—Cause, improper feeding, exclusive grain diet, lack of exercise. Not dangerous in itself, but frequently followed by prolapsus of the rectum, or what is commonly called piles. The constant straining causes this. The only remedy is laxative food and exercise. The protruding bowel must be washed clean as soon as seen and well covered with olive oil or lard. It should then be returned by applying firm pressure with the hand, and when once in place should be retained by three or more stitches of waxed linen or heavy silk thread, passed from side to side through the margins of the opening, care being used to take a deep hold in the skin.

While this operation is being done the animal should be held by the hind legs by two assistants, thus elevating the hind quarters. Allow stitches to remain two or three weeks.

RHEUMATISM.—A disease of the joints, manifested by pain, heat and lameness, with swelling of one or several joints. There may be high fever and loss of appetite. May be acute and rapid in its course, or slow, chronic and resulting in permanent enlargements of the bones of the legs, especially the knee and hock.

Causes.—Primarily deranged digestion, lack of exercise; dampness and exposure to draughts of cold air also a cause. The tendency to rheumatism is hereditary in certain families of hogs.

Treatment.—Endeavor to prevent by proper exercise, food and attention to surroundings. Do not breed rheumatic specimens even if fully recovered from lameness. In acute cases an adult hog should have twice or three times daily one drachm salicylate soda.

ASTHMA sometimes occurs in adult hogs.

Symptoms.—Shortness of breath on least exercise, noisy breathing, more or less intermittent. Do not breed; butcher early.

CONGESTION OF THE LUNGS sometimes occurs, the result of driving or chasing. May be rapidly fatal.

Symptoms.—Sudden shortness of breath and sudden great weakness. The hog is not adapted to rapid driving; if it must be driven at all, give plenty of time.

PNEUMONIA (LUNG FEVER) may follow congestion of the lungs; may be induced by crowding too many hogs together, when they heat and become moist, after which they are in poor condition to withstand cold.

Symptoms.—Loss of appetite, chills, short cough, quick breathing.

Treatment.—Separate sick at once from the drove; give dry quarters with abundance of dry bedding; tempt appetite with small quantities of varied food. Apply to sides of chest, enough to moisten the skin, twice daily, alcohol and turpentine equal parts; continue until skin becomes somewhat tender.

TETANUS (LOCK-JAW).—Caused by introduction into the system of the tetanus bacteria, which gains entrance through a wound.

Symptoms.—A stiffness of more or less the entire muscular

system, generally most marked in the jaws, which are greatly stiffened. Eating very slow, or entirely stopped; appetite not lost.

Treatment.—Some cases recover if carefully nursed. Give nourishing drinks, elevate trough or bucket so the patient can get its snout into the drink ; give dissolved in hot water and mixed with the slop forty grains bromide of potash two or three times daily until improvement is noticed. Do not attempt to drench. Any wound which seems to be a cause should be cleansed and wet often with five per cent. solution of carbolic acid and water.

LICE.—Very commonly found upon hogs. They are introduced by new purchases or by visiting animals.

Caution.—Examine the newly purchased hog well on this point before placing with the drove. Hog lice are quite large and easily detected on clean white animals, but not readily on dark or dirty skins.

Remedy.—Wash well with soap and warm water, if weather is not too cold, then apply enough petroleum and lard, equal parts, to give the skin a complete greasing. If weather is too cold for washing, clean with stiff brush. Creolin one part to water five parts is also a safe and sure remedy. Two or more applications are necessary at intervals of four or five days to complete the job. The woodwork of pens and rubbing places must be completely whitewashed.

MANGE.—Caused by a microscopic parasite which lives in the skin at the roots of the bristles.

Symptoms.—Intense itching with redness of the skin from the irritation of rubbing. Rather rare, but very contagious.

Treatment.—Separate diseased animals ; scrub them thoroughly with warm water and strong soap ; apply ointment composed of lard, one pound ; carbonate of potash, one ounce ; flor. sulphur, two ounces ; wash and re-apply every four days.

MAGGOTS.—The larvæ of the ordinary blow-fly frequently infest wounds on hogs during the summer months. Watch all wounds during hot weather ; keep them wet frequently with creolin one part and water six parts, or five per cent. watery solution carbolic acid. If the maggots gain entrance to the wound, apply either above remedies freely, or ordinary turpentine with a brush or common oil can.

ROUND WORMS.—Very common in shotes and young hogs, not apparently harmful, unless in great numbers, when they cause loss of flesh. They may be exterminated by keeping the hog without food for twenty-four hours, and giving to each shote or old pig one tablespoonful of turpentine thoroughly beaten up with one egg and one-half pint of milk. Good food and care will generally prevent serious injury from round worms. Hogs infested should not be pastured with others, or where others may pasture within a year. The adult worms are passed off with the manure, and being filled with eggs, render the pasture unsafe for many months, as the eggs withstand extreme and long continued exposure I lost several shotes, in fact eight out of a lot of nine. They would come to the trough and drink, apparently all right, then bound into the air, squeal and lie down and die. I was then told they had throat worms. I caught the only remaining one and poured down its throat a teaspoonful of spirits of turpentine. It squealed as loudly as any of them, but lived and raised a fine litter of pigs.

PARALYSIS OF THE HIND PARTS.—When hogs are affected by worms in the kidneys, they are sore across the loins and seem to have lost the use of their hind parts. When forced to do so they will get up and walk, but when the hinder parts are paralyzed they will not get up and cannot walk. For the last trouble, stimulate the surface with washing and rubbing with hot water, and keep the bed dry and clean. Turn them over and be patient ; they will generally get over it. They must have cooling and laxative foods. For worms in the kidneys, rub the back across the loins with spirits of turpentine every other day for a week, and if not better give a dose at the mouth on an empty stomach, one or two tablespoonfuls according to size. Do this two or three times. Dilute the turpentine with milk.

The most common form of tapeworm in man is derived from eating pork which contains the larval form of this parasite. The embryos are visible to the naked eye in infected pork ; each embryo is surrounded by a small bladder-like sac, about the size of a grain of shot. When such pork is eaten by man in uncooked, or partly cooked, condition, the embryo worms develop into adult worms, which reach many feet or yards in length. The mature worm in man is continually throwing off sections of its body filled with eggs. If these are eaten by the hog, they hatch in the hog's stomach and bore their way into the flesh of the pig.

Prevention.—Avoid the use of infected pork. Prevent hogs

having access to contents of water closets, or to land fertilized by the contents of water closets.

TRICHINÆ.—A disease of man due to eating pork containing trichinæ. Thorough cooking destroys the parasite, but infected meat is not safely used, and is condemned at slaughter-house inspection.

Prevention.—The feeding of hogs upon slaughter-house offal is a cause for spreading the parasites, and should not be practised. Rats are infected by eating slaughter-house offal, and as the rats are frequently eaten by the pig, infection likely often is the result. Exterminate the rats and do not feed offal.

TUBERCULOSIS (CONSUMPTION). A contagious disease common in man, cattle and not rare in the hog.

Symptoms.—Loss of flesh, cough, diarrhœa, swelling about the head and neck, which may open and discharge with little tendency to heal; death in from few weeks to months. Post mortem shows various sized tubercles, which may be situated in any part of the body, most commonly in the bowels, lungs, liver, or glands of the neck.

Causes.—Direct contagion from other hogs, but generally from feeding milk from tuberculous cows, or by eating butcher offal from such cows.

Prevention.—Care as to the source of the milk fed; if suspicious, boiling will render it safe. Do not feed butcher offal; separate suspicious hogs at once, and if satisfied they are tuberculous, kill and bury deep, or burn them. The tuberculin test can be applied to the remainder of drove, as without it it is impossible to say how many may be diseased.

WOUNDS generally heal readily in the hog if kept clean and free from maggots. The result of neglected castration wounds is sometimes serious. Have the animal clean as possible when castrated, and endeavor to keep it clean and give opportunity for abundant exercise until wound is healed. There is probably nothing better and safer to apply to wounds of the hog than creolin one part, water six parts.

TRAVEL SICKNESS.—Similar to ordinary sea-sickness in man; very common in shipping pigs by wagon.

Symptoms.—Vomiting, diarrhœa, great depression; seldom if ever fatal. May be rendered much less severe by very light feeding before shipment.

Swine, like human beings, suffer from wet feet.

The hog, unlike the farmer, grunts when grateful.

Don't breed " squealers ; " the well-bred hog is seldom noisy.

Bury the idea that anything is good enough to feed the hog.

Sows should be weeded out as well as cows. Keep only good milkers.

A tame pig will turn its owner a profit, a wild one is a nuisance on any farm.

Don't keep the boar with crooked legs, no matter what his pedigree may be.

Many old farmers scrub scabby pigs with buttermilk, and it proves to be a good thing.

The old legal fence in Pennsylvania was required to be horse high, bull strong and hog tight.

Kill a runt that won't grow with proper care, and in nine cases out of ten you will find traces of organic disease.

Sudden changes are usually to be avoided, but the change from a wet bed to a dry one cannot be too sudden.

Do not compel the brood sow to climb a steep plank to get into her pen : it causes serious injury and difficult births.

It has been determined by actual experiment that poor feeding is the great cause for extra development in the length of the snout.

All the improved breeds are able to equal their advertised performances, but it requires skill on the part of the breeder and feeder.

Here is a good way to lead a little pig. Tie the rope around his throat so it will not choke him, then carry it back and make a loop back of his legs. He can't get away if the rope doesn't break and you can hold him.

A friend of mine has a box with a slit at the bottom opening into a trough in which he keeps constantly a mixture of one pound of copperas, one pound of sulphur, one pound of black antimony, one-quarter pound of saltpetre, one quart of salt and one-half bushel of wood ashes. He has it in a dry place where all his hogs can get at it and thinks this is one reason why he has never had cholera on his place.

It used to be quite common in some sections to see pigs out to pasture and along the roads with yokes so they could not scramble through the fences. The yoke and manner of applying it are shown in the illustration. Now-a-days good wire fencing that is hog proof is so cheap and so universally used that I have not seen a yoked hog for a long time in my neighborhood.

Chapter XXII.

SUMMARY AND CONCLUSIONS.

Let the motto be, better pork at less cost.—John Tucker.

CHESTER WHITE.

To tell the weight of swine measure the girt in inches back of the shoulder, and the length in inches from the square of the rump to a point even with the point of the shoulder blade. Multiply the girt and length and divide the product by 144, multiply the result by eleven if the girt is less than three feet, or by sixteen if over three feet. The answer will be the number of pounds of pork. If the animal is lean and lank, a deduction of five per cent. from the above should be made.

Pork can be made better by feeding for quality rather than for quantity.

Pork can be made at less cost (far less than the average) by giving only the requisite amount of food, with muscle-making ingredients properly proportioned to fat-making ingredients.

These are the two lines along which farmers must seek increased profits in pork production.

It seems strange to say that skim-milk is really worth more for food than whole milk, and farmers do not generally so regard it. Yet such is a fact, provided it be fed in connection with corn or other carbonaceous food.   All farmers who are careless of skim-milk are wasting with every 100 pounds an article that is capable of producing twenty cents' worth of pork or ten cents' worth of manure.

Skim-milk should be fed sweet, it should be fed often (three or four times a day) and it should be fed warm.

I have a great many old and successful farmer neighbors, but find that most of them are still depending solely on the lessons of practical (and sometimes very costly) experience.   Few of them take the trouble to apply arithmetic to stock-feeding operations.   The younger ones, on the other hand, are on the alert, and are well read in recent farm literature, and these boys, as I call them, are sure to be heard from before many years.   They will be the leading farmers of the future ; not better men than their fathers, nor better citizens, but making money out of farming under conditions altogether different and vastly more scientific than prevailed a generation ago.

Of course, I cannot lay particular stress on the breed of hogs which happens to be my favorite, for my local surroundings are of course different from those of many other farmers who will read this book, and who have different conditions from those which have determined my choice.   I cheerfully admit that there are a half score of first-class breeds now well established in America, and if I should be compelled to change my home I might also be compelled to change my breed of hogs.

Cleanly, well-managed operations will result in pork of a superior quality, because of the better health and more rapid growth of the pigs ; and cleanliness will in all cases be accompanied by a large manure heap.

I give herewith the names and addresses of the various blooded swine breeders' associations, so that those desiring special information in regard to any particular breed may know where to get it.

American Berkshire Association. Secretary, Frank S. Springer, Springfield, Ill.

American Chester White Record Association. Secretary, Ernst Freigau, Columbus, O.

American Duroc Jersey Swine Breeders' Association. Secretary, T. B. Pearson, Thorntown, Ind.

American Essex Association, The. Secretary, F. M. Srout, McLean, Ill.

American Hampshire Swine Record Association, Secretary, E. C. Stone, Armstrong, Ill.

American Poland China Record Co. Secretary, W. M. McFadden, Union Stock Yards, Chicago, Ill.

American Tamworth Swine Record Association. Secretary, E. N. Ball, 1237 Volland street, Ann Arbor, Mich.

American Yorkshire Club. Secretary, Harry J. Krun, White Bear Lake, Minn.

Cheshire Swine Breeders' Association. Secretary, E. S. Hill, Freeville, N. Y.

International Ohio Improved Chester Record Association. Secretary, Herbert A. Jones, Himrods, N. Y.

Michigan Swine Breeders' Association. Secretary, E. N. Ball, 1237 Volland street, Ann Arbor, Mich.

National Duroc Jersey Record Association. Secretary, Robert J. Evans, 604 Main street, Peoria, Ill.

National Poland China Record Co. Secretary, A. M. Brown, Winchester, Ind.

O. I. C. Swine Breeders' Association. Secretary, J. C. Hiles, 30 Vincent street, Cleveland, O.

Ohio Swine Breeders' Association. Secretary, Ernst Freigau, Columbus, O.

Southwestern Poland China Record Association. Secretary, H. P. Wilson, Gadsden, Tenn.

Standard Chester White Record Association. Secretary, W. H. Morris, Indianapolis, Ind.

Standard Poland China Record Association. Secretary, George F. Woodworth, Maryville, Mo.

Victoria Swine Breeders' Association. Secretary, Geo. Davis, Dyer, Ind.

INDEX.